Postharvest Losses, Technology, and Employment

About the Book and Author

This study of Bangladeshi rice farmers examines the proposition that food availability can be increased cost-effectively by technological change to reduce farm-level postharvest food losses. Dr. Greeley argues that food losses in traditional postharvest systems are considerably lower than is usually assumed. A social benefit-cost analysis of the two major technological innovations in postharvest rice processing - pedal threshers and rice mills - shows that there is an adverse income distribution effect associated with the new machines, but only in the case of rice mills do these welfare losses exceed the production gains. The primary losers in the process of technological change are female wage labourers from the poorest rural households. In practice, policymakers have limited capacity to control the diffusion of this technology; there is a need for innovative approaches to income-generating programs attuned specifically to these female wage laborers as well as to others adversely affected by the new technology.

Martin Greeley, an economist, is a Fellow of the Institute of Development Studies at the University of Sussex.

Postharvest Losses, Technology, and Employment

The Case of Rice in Bangladesh

Martin Greeley

LONDON AND NEW YORK

First published 1987 by Westview Press, Inc.

Published 2019 by Routledge
52 Vanderbilt Avenue, New York, NY 10017
2 Park Square, Milton Park, Abingdon, Oxon OX14 4RN

Routledge is an imprint of the Taylor & Francis Group, an informa business

Copyright © 1987 Taylor & Francis

All rights reserved. No part of this book may be reprinted or reproduced or utilised in any form or by any electronic, mechanical, or other means, now known or hereafter invented, including photocopying and recording, or in any information storage or retrieval system, without permission in writing from the publishers.

Notice:
Product or corporate names may be trademarks or registered trademarks, and are used only for identification and explanation without intent to infringe.

Library of Congress Cataloging-in-Publication Data
Greeley, Martin.
Postharvest losses, technology, and employment.
(Westview special studies in social, political, and economic development)
Bibliography: p.
1. Rice trade--Bangladesh. 2. Rice--Bangladesh--Processing--Cost effectiveness. 3. Women rice workers--Bangladesh. I. Title. II. Title: Post-harvest losses, technology, and employment. III. Series: Westview special studies in social, political, and economic development.
HD9066.B32G73 1987 338.1'7318'095492 87-10648
ISBN 978-0-367-28397-1 (hbk)

ISBN 978-0-367-29943-9 (pbk)

Contents

Tables and Figures

TABLES

FIGURES

Acknowledgments

I gratefully acknowledge the financial support of the British Overseas Development Administration, without which this book would not have been possible.

I am deeply indebted to a large number of people who have helped me in a variety of ways since this study was first proposed in 1977. I am very grateful to all of them and, whilst space does not permit me to acknowledge every individual, I do wish to record my very special thanks to a few without whom this study could not have been completed.

Whilst in Dhaka more than twenty colleagues and friends worked with me on the IDS - BCSIR Postharvest Project and it is their collective endeavour which has provided the opportunity to write this book. I am especially grateful to Dr. S F Rubbi, Director of the Food Science and Technology division, BCSIR, for his unceasing willingness to accommodate all the demands I made on the resources of his institute.

I am very grateful to Robin Boxall who, over the course of the last ten years, has contributed enormously to my understanding of the technical and scientific side of postharvest food systems.

Several secretaries have been tortured by my appalling handwriting and I am especially grateful to Marion Huxley, Susan Saunders and Linda Simmons who have all worked with great energy and dedication to produce this book.

My biggest debt is to Saleha Begum who has stoically borne the brunt of the domestic disruption associated with writing the book and who has also provided me with many insights concerning the rural economy of Bangladesh.

Finally, it is a pleasure to express my deepest appreciation to Michael Lipton for first introducing me to this subject area and for innumerable discussions which have contributed enormously to my understanding of the issues involved.

Martin Greeley

1

Outline of the Study

1.1 INTRODUCTION

Insufficient food production and inequitable food distribution are the two main themes of commentaries upon the problem of hunger in developing countries. In the last decade there has been increasing awareness of an alleged third option - prevention of postharvest food losses. The possibilities of increasing food availability by reducing wastage after harvesting have been receiving increasingly widespread interest from both governments and aid donor agencies. Attention has been focused on the dietary staples especially the major cereals, maize, rice and wheat.

All postharvest operations up to threshing take place at farm-level and, since the major part of food production in developing countries remains on the farm that produced it, subsequent operations such as storage are often farm-level activities. The prevention of postharvest food losses at farm-level has therefore been singled out as the main target for intervention programmes.

This book presents a Bangladesh case study of the farm-level postharvest system. There are two main objectives. First, to use measured estimates of food loss to test (and reject) the conventional assumptions: that postharvest farm-level food losses are large; that they can be prevented cost-effectively by technical change; and that as a consequence, there will be more food consumption by hungry people. Commonly, none of these assumptions are true and the evidence from Bangladesh, plus supporting evidence from elsewhere, is used to show why they are wrong.

In doing so we do not deny the crucial policy significance of postharvest technical change. Postharvest farm-level technical change currently occurring in Bangladesh is reducing food availability, reducing the number of work places and is particularly damaging to the welfare of some of the very poorest rural households. The second main objective is to explain farmer adoption of these technical changes in terms of cost savings; to explore the social costs and benefits that they give rise to; and, to draw policy conclusions on the types of intervention required. Evidence is reviewed concerning the two major farm-level postharvest technical changes now occurring, the introduction of pedal threshers and of huller rice mills. The rest of this short introduction provides a brief description of the data sources for the study and an outline of the arguments in each chapter.

Most of the empirical material presented in this study derives from a three year research project (1978-1981) 'Rice in Bangladesh: Appropriate Technology for the Intra-village Postharvest System' which was undertaken by the author in Bangladesh in collaboration with the Bangladesh Council of Scientific and Industrial Research. This project, described more fully in section 4.2, generated three main sorts of data: a) measured estimates of food loss in farm-level postharvest operations; b) economic, including food loss, data on the two new techniques, pedal threshers and huller rice mills, currently being introduced into Bangladesh; and c) earnings data over a year, collected on a weekly recall basis, from a hundred of the poorest households in the project field areas; approximately fifty per cent of the sample were from households where the adult women worked as wage labourers and the focus of the data collection exercise was on the sources of female earnings and the size of their contribution to households. These three sets of data were collected by a team of graduate men and women who lived for two years in two project field sites (near Madhupur, Tangail district and Chandina, Comilla district). This data is used in Chapters Four to Seven; in addition there was a preliminary village survey, covering all 3094 households in the eight villages, detailing economic and postharvest data which is used from Chapter Three onwards to describe particular features of economic conditions in the villages studied.

1.2 CHAPTER OUTLINE

Chapter Two has two main objectives. First, to describe and explain the recent development of policy interest in postharvest food loss prevention. Secondly, to outline economic considerations, both from a private and a social viewpoint, that affect choice of postharvest techniques but which have not always been incorporated in the loss prevention debate.

It is argued that the advocates of loss prevention programmes were aware of the unreliability of available estimates but nevertheless used them, often selectively, to support their case. Technological determinism, the impact of the green revolution, the pressure from food donors and commercial pressures are four influences identified as important in developing support for food loss prevention programmes.

The consequence of this concern with loss prevention was a narrow concentration upon the food loss characteristics of postharvest techniques; the second part of Chapter Two builds on the choice of techniques literature to explore the economic considerations other than loss prevention which affect the private and social profitability of postharvest technical change. Ten 'farmer' considerations are described which, it is argued, make the assessment of loss prevention interventions complex and serve to illustrate the limitations of a loss-prevention focus in considering postharvest technical change. Finally, the principles of social benefit cost analysis (SBCA) are outlined briefly to examine the relationship between maximising aggregate consumption - the objective function in SBCA - and reducing postharvest food losses.

Chapter Three provides the Bangladesh context for the analysis in subsequent chapters. First of all it provides working definitions of postharvest and food loss and argues that the focus of loss assessment studies should be on quantitative losses. Measurability, sound definition and economic valuation are identified as the all-important aspects of a postharvest food loss estimate. It is suggested a) that a single comprehensive valuation of loss from all causes, quantitative as well as qualitative, is not possible; and b) that valuation exercises ought to take account of the stage in postharvest operations that losses occur - this is not usually done.

Section 3.3 outlines cropping patterns and traditional postharvest operations in Bangladesh and identifies features of importance to food loss assessment exercises. The chapter then examines the objectives of current national economic planning in Bangladesh and goes on to relate the two principal planning objectives, food self-sufficiency and rural employment generation, to postharvest technology issues.

The next section (3.5) argues that the predominant emphasis in the Bangladesh postharvest literature has been on new technology to prevent food losses, in exactly the same way as the more general postharvest literature reviewed in Chapter Two; it is shown that this 'high food loss lobby' has not been supported by field experience and that no projects designed to introduce loss-reducing techniques to farmers have been successful. The review concludes with a discussion of the remaining (non-food loss) Bangladesh literature related to postharvest technology, which in various ways identifies the need for the type of research reported here.

The final section of Chapter Three on product, capital and labour use employs survey data to describe the pattern of labour and capital use in traditional postharvest operations, which are shown to be very labour intensive, and contains an extended discussion on rice end uses. Postharvest studies often do not distinguish between the value of a unit loss of grain for on-farm consumption and grain for sale off the farm; reasons are given for believing that with qualitative losses a distinction exists and should be made, especially in view of the increasing emphasis being given by postharvest specialists to qualitative deterioration.

The main evidence on food losses is presented in Chapter Four. One of the criticisms of past research made in Chapter Two is the failure to provide full descriptions of methods employed to measure food losses and one of the reasons for generally inadequate evidence on the size and distribution of losses has been the inadequacy of applied loss assessment methods; therefore, after a brief overall description of the Bangladesh postharvest project, three sections of Chapter Four are devoted to a discussion of the loss assessment methods employed in the project for each postharvest operation. Project work involved experimentation with different loss assessment methods for some operations, eg cutting to threshing, to establish the best method and this work is described to illustrate the very real practical problems of generating reliable figures on farm-level food losses.

The best estimate of quantitative losses across all farm-level postharvest operations is seven per cent; section 4.6 examines the breakdown of losses by operation and argues that even for the storage operation, where losses are highest, cost-effective loss prevention intervention is ruled out by the low level of losses. Section 4.7 looks at the influence of seasons, regions, farm-size and varieties on loss-levels and, with some limited qualifications, argues that these four factors do not change the conclusion that loss prevention intervention is not cost-effective.

A major concern in Bangladesh postharvest studies has been with moisture-induced qualitative deterioration and section 4.8 gives a detailed account of the wet season studies undertaken as a part of the project. The results show that wet season qualitative deterioration only leads to quantitative loss under exceptional circumstances and, for the major share of production which is kept at farm-level, the wet season is not an important cause of economic loss to farmers.

The evidence on mycotoxins presented in section 4.9 does not indicate serious levels of toxicity, though results are only from a preliminary study; however, the incidence of the toxin-producing microflora supports the wet season study in identifying extended pre-threshing stacking as the most problematic operation.

Overall, it is concluded (section 4.10) from the evidence on losses that there are very limited opportunities for future research on farm-level loss prevention intervention. The marketed surplus during the wet season is tentatively identified as a focus if such research is funded but it is suggested that new institutional arrangements are necessary in the organisation of research. The final section briefly presents supporting evidence from eight other countries of recent loss assessment studies which have measured losses and also found them to be low.

Given the evidence on low food losses, it is apparent that loss prevention is not the motivation for farmer adoption of new techniques. Postharvest innovation cannot, in the absence of food losses, be output enhancing and therefore must be cost-reducing; and, since traditional postharvest techniques are very labour intensive such innovation is inevitably labour-displacing. Chapters Five and Six examine the two technical changes now occurring in postharvest operations, the introduction of pedal threshers and of huller rice mills, and pay particular attention to the labour use characteristics of the new techniques compared to traditional methods.

Chapter Five shows that despite the fact that pedal threshers cause higher food losses, they are still a profitable investment both for the machine-owner and for farmers renting-in pedal threshers. There is a substantial reduction in labour hours per ton as a result of adopting the pedal thresher but the interesting feature of the pattern of labour use is that it is all family labour that is displaced. Three types of explanation are offered: an absolute supply constraint of family labour per ton, as yields increase; elements of monopoly control over labour supply by machine-owners; and farm management constraints on family labour.

This last argument is discussed at some length since the implication made, that family labour has a higher opportunity cost because of its farm management function than wage labour, is contrary to perceived wisdom (and a lot of evidence) concerning the relative valuation of labour types. There is insufficient evidence to reach a firm conclusion on this issue but the importance of farm management, usually neglected in farm budget studies, can be expected to increase as cropping intensity increases, or yields increase due to higher input use and improved timeliness of operations; it is for these reasons that the farm management function of family labour is offered as a partial explanation of the unexpected pedal thresher labour displacement pattern.

Without providing a formal SBCA, section 5.7 considers how the use of efficiency and social prices would affect the profitability of the investment and whether, given the effects on income distribution, there is a net social gain from the innovation. Whilst it is not possible to give unequivocal support for the diffusion of pedal threshers, because of ambiguity concerning distributional implications, it is argued that pedal threshers are very close to what could be considered 'appropriate' technology for Bangladesh. Their varietal, and therefore to some extent seasonal, specificity and the fact that other threshing innovations, with clearly negative effects on aggregate social welfare, are rendered less imminent are two significant features of pedal thresher use which strengthen this conclusion.

Chapter Six examines the second technical change - the displacement of _dheki_ use (manual rice husking) by the huller rice mill. The analysis is concerned very specifically with this labour displacement in relation to milling of farm-level consumption paddy since the marketed surplus is almost all mill processed already and the technical change now occurring relates to this consumption paddy. The two techniques are described and data on their distribution and labour use pattern is presented (sections 6.2 and 6.3). There is considerable uncertainty about the share of mills in processing of farm-level consumption paddy and, after reviewing the evidence, it is estimated that

25-35 per cent of on-farm consumption paddy was mill processed at the time of the survey.

Section 6.4 argues that despite the differences in outturn between the dheki and the mill there are ultimately no differences in food availability associated with the choice of techniques because the lower outturn of the huller mill is recovered by family farm women in cleaning and separating the by-products. Section 6.5 examines under what conditions the adoption of the mill is profitable to the farm household and section 6.6 establishes, predictably, the private profitability of mill ownership.

Sections 6.7 to 6.11 present a fairly detailed social benefit-cost analysis of the choice of techniques. The change in technique results in substantial labour saving, largely of family female labour but female wage labourers are also displaced with serious consequences for their earning capacity and for the welfare of their households. The critical issue in the SBCA is to establish whether the social value of the welfare loss suffered by this group is sufficiently large to outweigh the benefits derived by farmers and mill owners. Ideally, a set of accounting prices and distributional weights would be available to make this assessment since SBCA as a planning tool is only effective when a consistent set of values are used in all appraisal exercises. However, such values are not available and the accounting prices are derived in sections 6.9 and 6.10. The critical values, which are described at some length, are the accounting price for labour and the income distribution weights; accounting prices are used for other inputs and for the output (milling services) but their derivation is fairly crude (in the right direction rather than rigorously estimated since no satisfactory basis for improved estimates is available and results are insensitive to their value).

The results of the comparison are presented first at market prices, then at efficiency prices and finally at social prices. They show that the huller mill is much more profitable than the dheki until distributional weights are introduced; at this, the social price stage, the dheki is very much more profitable. This result is highly robust and not affected by even quite large changes in assumptions about the values of specific parameters.

If distributional weights are accepted as a part of choice of techniques analysis, as it is argued they should be, then the results establish clearly the social profitability of keeping the dheki. The only reason why this reversal occurs between efficiency and social prices is the deep poverty of the female wage labourers displaced by the mill. The output foregone elsewhere by employing them in dheki work is very small and the social value of their consumption is very large so there is a net social benefit by employing these women in dheki work.

These results provide the point of departure for Chapter Seven on the policy implications of the research. First, it is argued on two related grounds, regulatory impracticability and limits on planners' effective control, that the results of the SBCA do not necessarily indicate a policy of stopping huller mills - programmes designed to benefit the displaced women are the alternative and probably more effective policy response. It is suggested that: a) in this respect the situation in Bangladesh may be typical of other countries because links between farm-level postharvest technical change and women's welfare are almost universal; and b) that to develop programmes to benefit this displaced female labour it is necessary to understand the broader socio-economic context in which this technical change is occurring.

This is done by examining first the gender division of labour in Bangladesh (section 7.2) and secondly the process of peasant differentiation and its effect upon female work (7.3). The argument developed, based on these two sections, is that the economic role and social status of women from farm families is significantly different to the displaced wage labour women who are mainly from landless families; the differences mean that effective policy responses to the needs of these two groups in the form of intervention programmes are substantially different. It is also argued that development programmes for wage labour women provide a unique opportunity to address both gender issues and poverty issues effectively. The chapter concludes with a number of features, relating to programmes for wage labour women, that recent field experience has shown to be important.

2

The Postharvest System and World Hunger: The Debate on Neglect and the Neglects of the Debate

2.1 INTRODUCTION

This chapter examines the conceptions and misconceptions concerning the role of postharvest food loss prevention in the fight against hunger. Section 2.2 traces the development of postharvest food loss prevention as an issue in food policy. Section 2.3 looks at some of the main evidence used in the debate on food losses and section 2.4 examines the main influences that determined attitudes towards the prospects for increasing food availability by reducing postharvest losses. It is argued that the overwhelming emphasis on food loss prevention as the motivation for adopting new postharvest technology resulted in the almost universal neglect of more thorough and systematic economic and social analysis of planned postharvest intervention. This argument is developed in section 2.5 on the neglects of the debate and section 2.6 provides a discussion of the costs and benefits of loss-reducing technical change. This discussion is concerned first with the range of factors, other than food loss, that will influence the farm-level investment decision on postharvest technology. It then goes on to consider the principles underlying social benefit-cost analysis and how their incorporation in appraisal of planned intervention might affect the choice of postharvest technology. Finally, there is a discussion of employment and food self-sufficiency objectives in relation to postharvest technological choice.

2.2 THE DEBATE ON NEGLECT

The postharvest system has been described as 'the neglected dimension in increasing the world's food supply' (Bourne 1977). Fuelled by this and similar descriptions (FAO 1975, Spurgeon 1976) there was a growing concern during the 1970s with the potential contribution that prevention of food loss could make to the reduction of world hunger. By 'attacking a tragic waste' (Ceres No.60 1977, p.12) it was widely assumed that food availability could be substantially increased. Advocates of postharvest food loss prevention (PHLP) programmes used various numbers to support their case: it was claimed that 168 million people could be fed (NAS 1978, p.7), that the value of reducing losses by half was $7.5 thousand million and that food saved would match the volume of food aid (Kissinger 1975 cited in Bourne 1977, p.3).

Another stratagem was to compare alleged levels of postharvest losses with production deficits. Parpia (1977 p.20) for example argued that: 'In most of the food-deficit countries, actual shortages represent 4-6 percent of requirements, while losses have been estimated at 20 to 40 percent of production'. The presumption that postharvest food losses were large in relation to production deficits was a powerful argument for investment in PHLP. National governments in developing countries and aid donors came under increasing pressure to support postharvest research and development activities as a knowledge of this alleged 'neglected dimension' percolated through the development community. The World Food Conference of 1974 devoted special attention to it1/ and in 1975 the seventh special session of the UN General Assembly passed a resolution that 'The further reduction of postharvest food losses in developing countries should be undertaken as a matter of priority, with a view to reaching at least a 50 per cent reduction by 1985' (Bourne 1977 p.3).

Initially at least there were few attempts to be precise about the exact level of food losses in particular postharvest systems, about the costs of PHLP or about the distributive effects of postharvest technical change. In the absence of evidence to the contrary the appeal of food loss prevention as a solution to problems of world hunger was very strong. There were three key unquestioned assumptions: that food losses were high; that they could be prevented cost-effectively; and that there would be less hunger as a consequence.

These assumptions were responsible for the growing concern amongst development and food policy groups both nationally and internationally to develop PHLP programmes. The National Academy of Sciences (1978) produced a special report, Post-Harvest Food Losses in Developing Countries. USAID in collaboration with ODA and FAO produced a manual (Harris and Lindblad 1978) on food loss assessment methods for developing countries. These and other initiatives on PHLP, were the initial response to the realisation of neglect. It was a period of difficulty for postharvest scientists who were often well aware that the real neglect was in the accurate estimation of food losses but who seemed unwilling to question the 'high food loss' assumptions that were the source of their increasing attraction to funding bodies. As Bourne (1977, p.15) stated 'There is often the temptation to cite "worst case" figures to dramatize the problem ... In some cases there is the temptation to exaggerate the figures of loss particularly if there is a prospect that high figures of loss will prompt aid or grants from some donor'.

There was of course a wide range of opinion on the real potential of PHLP. This section presents the extreme view of the debate which remains the popular view and which provided the initial impetus. It was derived principally from a belief that a very substantial contribution could be quickly and cost-effectively made to reduction of hunger through farm-level PHLP. This belief had been endorsed as early as 1968 in the annual FAO survey 'The State of Food and Agriculture' when farm-level storage loss reduction was identified as one of five areas for emphasis in future programmes. However, it was not until the 1974 World Food conference that international funding to reduce postharvest food losses became a focal issue (Bourne 1977, p.2). By the late 1970s most of the main donor agencies, including AID, EEC, FAO, GTZ, IBRD, IDA, IDRC, ODA and SIDA were funding postharvest food loss prevention activities in collaboration with national governments.

FAO played an important encouraging and co-ordinating role, and bilateral agencies such as ODA, with a history of postharvest involvement and with specialised institutions such as the Tropical Products Institute, were key participants in the development of postharvest aid pogrammes.2/ There was rapid growth in levels of funding. Between 1976 and 1978 AID raised its authorised expenditure for postharvest related programmes from $1.7 million to $8.5 million (AID 1977) and in the same period increased its loans (chiefly to Latin America) for similar programmes from

$4.8 million to $14.9 million.3/ FAO authorised expenditure in 1976/77 was $2.5 million (FAO 1975, Annex I) and by 1981 it was over $19 million with a further $15.6 million of project requests under consideration (Nicholas 1981, p.3).

The impetus for the development of PHLP activities was provided by the belief that high food losses were an almost universal phenomenon on the farms of developing countries. The harvest failures of the early 1970s4/ and the limited regional impact of the HYV seeds in reducing production deficits encouraged new avenues of research. Support for research and development on loss preventing technical change in the postharvest system was based on the contribution such change could make to solving problems of hunger. Despite the large number of studies on use of HYVs which had underlined the importance of analysing the income distribution effects of technical change, the advocates of farm-level postharvest technical change to prevent food losses paid little attention to such issues. Whilst some authoritative commentators made cautionary remarks about 'the lack of appreciation of the cost-benefit relationship attached to the various procedures for postharvest loss-reduction', (Spensley 1977, p.18) the more influential attitude is aptly summarised by the FAO view (1975, p.7) that it is '... obvious even to the lay observer that action is necessary'.

A characteristic idiosyncrasy of contributions to the debate on losses was the juxtaposition of a) the recognition that loss assessment methods were inadequate and that research was needed to improve the numbers being used with b) the view that losses were high and that PHLP was necessary. A possible explanation is the fact that the impetus for funding of postharvest programmes arose from popular concern with saving food and feeding people and not from professional concern with perfecting understanding of the problem. Whilst enormous progress has been made in the decade since the 1974 World Food Conference in food loss research methods, the major part of postharvest funding has been for PHLP programmes without prior research. For example, it was only in 1983 that FAO began its first comprehensive loss assessment research programme at farm-level. It is only in the last two or three years that the frequent failure of PHLP programmes to convince farmers to change techniques has led to serious questioning of the rates of return to loss prevention. Whilst 'in-house' opinion in influential bodies such as TPI and FAO has certainly modified - indeed TPI research, and more recently FAO's, has helped reform opinion - the popular view remains that food availability can be increased, and cheaply, by PHLP programmes.

2.3 THE EVIDENCE ON FOOD LOSSES

Whilst there were a large number of '<u>guestimates</u>' to the general effect that food losses were 20 per cent or more, there were very few actual field estimates. Moreover, most estimates that did purport to be based upon field research gave no details of their research methods. Statements such as 'postharvest losses are 20 per cent' beg at least three substantive issues: the definition of postharvest, the definition of loss and the accuracy of measurement techniques. These issues are discussed in more detail in chapter 3 and here we only wish to make the point that specific estimates were often of uncertain worth because: which postharvest operations were included and which ones were not was unclear; the definition of loss was not given and therefore the value of such losses, to the owner of the crop, to the consumer or to society could not be estimated; and the accuracy or comprehensiveness of the measure of loss could not be judged because detailed methods of estimation were not given.

Although development of a commonly accepted replicable loss assessment methodology was recognised as a priority by postharvest specialists (e.g. FAO 1975, p.5), the estimates of loss that were published were rarely discredited for their methodological inexactitude. On the contrary, they were used to promote support for PHLP programmes that sometimes included loss assessment as one element but generally were designed either to promote PHLP extension services or to develop new postharvest equipment that would reduce food loss levels. This point is illustrated below with an example from India before turning to the most well-known set of loss figures.

Indian Assessment Of Loss Prevention Prospects

One of the earliest sets of Indian postharvest loss figures was first published in 1965 (cited in Parpia 1977, p.21) by the Central Food Technological Research Institute in Mysore which gave all-India all-grains loss figures of 60 per cent across all operations. This figure based upon 'expert' guesses was broken down into field loss 25 per cent, storage loss 15 per cent, handling and processing loss 17 per cent and other losses 3 per cent. It seems unlikely that many policy-makers seriously believed these figures. This point was succinctly made by an USDA attache in New Delhi, quoted in Brown (1970), who collated various local estimates of operation or cause-specific loss that together gave total postharvest losses of 105 per cent! But if losses were even one tenth of this level, and could be saved cost-effectively, the savings would make a substantial contribution to solving Indian food problems - at least as far as physical availability was concerned. In other words, figures like these, if not to be taken literally, nevertheless strongly supported the case for increased financial support for PHLP.

The Government of India in fact appointed an Expert Committee to review the evidence on food losses in an effort to resolve the uncertainty surrounding what loss levels were and to identify PHLP possibilities. A much quoted Interim Report (Government of India 1967) gave total losses of 9.33 per cent (not 60%!) but the Final Report (Government of India 1971) concluded that the evidence to support this or any other loss estimate was woefully inadequate and the Committee felt unable to support any statement about losses until more reliable loss estimation exercises had been undertaken. This report was not published - perhaps because it did not support the arguments then being made in the Department of Food for increased funding of PHLP activities.5/

The Indian case illustrates the scant importance attached to scientific objectivity in stating the arguments for PHLP interventions. It has tended to be literature reviews by eminent research bodies such as the Administrative Staff College (1976, as part of an optimal store location study) and the Birla Institute (1979) that are most widely publicised. These reviews have not always been sensitive to the accuracy of available figures and more generally the problem of defining losses meaningfully. Much Indian scientific research on postharvest technology has therefore gone straight into the business of developing improved postharvest techniques and has not attempted to measure food loss during traditional practices.

The few studies that did measure food losses appeared to be ignored. Analysis of the attitudes in 35 papers presented to a 1981 Government of India organised workshop6/ in New Delhi on postharvest technology showed that there were 5 papers reporting grain storage loss levels. All of them reported farm-level loss levels of 5 per cent or below. Of the other 30 papers, all contained phrases like 'a national problem of great neglect', 'a quarter of our production is lost', etc. The loss figures available were not maligned merely ignored - the existence of a 'neglected dimension' and the opportunities it allegedly offered were not regarded as being at issue. The Indian scientists and researchers had, still have, more and better data in relation to postharvest losses than other developing countries. The problem faced there as almost everywhere else was essentially a cognitive problem; the evidence on losses was presumed to support loss-reducing technical change and the perceived policy issue was planning intervention to reduce losses.

Influential Figures And Their Limitations

There were particular sets of figures that were reproduced several times and were more influential than others in giving a priority to PHLP. The most widely quoted food loss figures have been variously attributed to de Padua (eg in Bourne 1977 and USDA 1980) and to FAO (e.g. in Parpia 1977 and Bergeret 1981).7/

The figures (de Padua 1976, p.6) are for rice and read:

Harvesting	1-3 per cent
Handling	2-7 per cent
Threshing	2-6 per cent
Drying	1-5 per cent
Storage	2-6 per cent
Milling	2-10 per cent

These have been quoted as applying to the Philippines and also to the whole of South East Asia. The first four estimates were in fact prepared by Samson and Duff and presented in a seminar at IRRI in July 1973. As Russell (1980) describes, their base is actually quite narrow: 'for the first four categories, Samson and Duff, on the basis of an experimental trial conducted at the International Rice Research Institute in the 1972 dry season and field surveys in Central Luzon, Philippines, in the 1972 wet season and 1973 dry season, estimated ...' Russell continues 'subsequently de Padua added two more categories of estimates "based on collective measurements done by participants in postharvest technology training courses over a period of five years"'. These estimates have figured prominently in many major postharvest loss prevention policy review papers produced by the donor agencies, eg USDA (1978), Spurgeon (1976), and Bourne (1977), and have undoubtedly been highly influential in promoting increased funding for postharvest research and development. They have had an attraction because they appear to be comprehensive and because, however spurious, their form of presentation implies a degree of accuracy in their estimation. Russell discusses seven different problems with them relating to their presentation and interpretation. These largely relate to omissions, possibilities of double counting and interaction between operations such that achieving low losses at one stage, eg field transportation, can increase losses at subsequent stages, eg threshing.

One fundamental problem, which Russell and others have pointed out, is that the estimates are non-additive as the amount of grain at the start of each operation is different. For example, after a harvesting loss of 3 per cent only 97 kilograms out of an original 100 kilograms would enter the handling stage. A 7 per cent loss in handling from this 97 kilograms would be 6.8 kilograms or 6.8 per cent of the 100 kilograms originally available. These figures have been cited (Spurgeon 1976, Parpia 1977) as showing a maximum total loss of 37 per cent but allowing for non-additivity the true total maximum loss is 32 per cent.

A related problem lies in the derivation of operation-specific loss percentages. The adjustments for non-additivity given above assume the per cent loss figure was calculated on the amount of grain entering into each operation. In fact, researchers often estimate percentage loss using the grain remaining <u>after</u> the operation as their base. This is an important difference; for example, if one hundred units of food are produced the physical loss of twenty units is one quarter of the remaining 80 units but only one fifth of the original 100 units of food produced. It is not stated which method was followed and the presentation of the original tables for the first four estimates in Samson and Duff (1973) does not allow the reader to work back and derive the basis with a high degree of certainty; however, it appears from the description of sample collection methods and the figures on field yields and field losses (Samson and Duff 1973 Tables 4 and 5) that the harvesting figures, at least, used the wrong basis - i.e. yield was net of losses. Unfortunately, use of grain remaining <u>after</u> losses have occurred has now been recommended as the basis for estimating percentage loss for all FAO loss assessment work (FAO, 1981b) - so it is likely, if FAO's methods are accepted widely, that most future estimates will be overestimates - and the larger the estimate the larger will be the amount of overestimation.

The biggest drawback of the harvesting, handling, threshing and drying estimates however is that they are controlled experiments and not based on farmer practice. An example will clarify the significance of this. The handling losses (2% to 7% and the highest of the four estimates prepared by Samson and Duff) were based on experiments in the 1972 dry season on the IRRI farm; the objective was to ascertain the effect of harvesting date (days after seeding) on postharvest losses. The 2 per cent loss occurred with harvesting 113 DS (days after seeding) and the 7 per cent loss with 125 DS (Samson and Duff 1973, Table 3). Each figure was based on just three replications. Without information on farmer practice it is not possible to make any statement about the relevance of these experimental farm results to actual farm-level losses. This information is not given.

The field experiments in the two following seasons were similarly organised with crop cuts being taken before, at and after the crop maturity date. These appear to have been the only source (it is not absolutely clear from the paper) of the harvesting and threshing loss figures and these results therefore suffer from the same problem of interpretation. Thus, of the first four (Samson and Duff) estimates, three of them (harvesting, handling and threshing) cannot be interpreted, without more information, to yield estimates of farm-level losses.

Only the drying results could be safely interpreted as evidence of the losses occurring due to current farmer practice compared to improved practice. This is because they are based on a comparison between paddy that is sun-dried and paddy which is mechanically-dried for the same harvesting date. The traditional practice is sun-drying and the table states that this practice causes losses of 1-5%. These results suffer another problem though. It is unclear from the tables (Samson and Duff 1973) what measure has been used to estimate drying loss. There are three possibilities; <u>milling yield</u> (the percentage rate at which paddy, rough rice, is converted into polished rice), <u>rice yield</u> (kilograms per hectare - to use this requires carefully designed research); and <u>physical losses</u> in drying. The paper provides data on the first two but not on the final possibility. The first index of loss in drying is total <u>milling yield</u> (rice outturn as a percentage of paddy input by weight); for this index traditional sun drying performed better in 20 out of 38 cases and in 5 the performance was the same. These results could not support a 1-5 per cent loss estimate. (The mechanical drier produced more head rice (whole grains) in all but one case and the most extreme difference was 22 per cent; but head rice yield is not total milling yield and therefore cannot, alone, be a measure of loss.) The only other data presented in relation to drying method is total milled rice in kilograms per hectare. However, in 12 out of 22 of these rice yield results sun-drying has a higher yield than mechanical drying and in another five they are the same. Again, these results could not be the basis for the 1-5 per cent loss figures. Finally, if the basis for the loss estimate was physical losses during drying, no information is provided on how the results were derived.

The preceding few paragraphs are not intended as criticism of Samson and Duff. Their study design allowed them to relate harvesting date to various types of postharvest loss. They have done this and their results support other studies in establishing that the optimum harvesting date is the approximate date of maturity, when the moisture content is around 20 per cent. On this point their results are quite clear and the only confusion the reader faces, as we have seen, is in interpreting the drying results. The misuse of their results, which is not the authors' faults, has been in citing them out of context as evidence of what farmers are actually losing. The 7 per cent upper boundary for handling loss for example would occur only when farmers cut their crop nearly two weeks after maturity. It also presumes that even if cutting is delayed two weeks farmers would not reduce the three weeks of field and bundle drying between cutting and threshing; one reason for delaying cutting is to reduce drying time so it is much more likely that as a consequence farmers would reduce field drying time. Moreover, higher handling losses when harvest is delayed are largely due to shattering and are likely to be associated with lower than normal threshing losses as threshing becomes easier. This is not taken into account in the results.

No information, other than the basic source given above, is available about the last two estimates, for storage (2-6 per cent) and milling (2-10 per cent) provided by de Padua. It is worth noting though that milling losses, unless carefully measured, will actually reflect damage at earlier stages especially drying. Since the drying and milling results reported here are from independent sources there is every possibility of double-counting; ie that reduced milling yields due to poor drying are included as both drying losses and milling losses.

Other than Russell (1980) none of the people using these figures has made any reference to their inherent and serious limitations as evidence on farm-level food losses. If farmer practice corresponds most closely to the experimental conditions which gave rise to the lower boundaries of the loss ranges the conclusion to be drawn is that in no operation does the farmer suffer losses of more than two per cent. This is hardly evidence of cost-effective loss-prevention intervention opportunities. If farmer practice does not correspond closely to the experimental conditions giving low losses, the first question to ask is why not, when the means (harvesting at maturity) to do so are available. These figures could tell us a great deal if we knew what farmer practice actually was and the economic basis for his decisions about his practices. Taken alone, they tell us very little and, given their widespread use, most rank as an example of the problem Bourne (1977, p.15) describes where 'figures that have been obtained by careful measurement are manipulated for various reasons'. These figures continue to be quoted (ASEAN, October 1985) as evidence of up to 37 per cent losses at the farm-level in South East Asia.

Turning from specific figures to the range of figures that informed the debate on losses, Table 2.1 gives 28 estimates of losses up to 1978 for rice which vary between 1 and 40 per cent. It has been reproduced from a widely quoted and influential report of the United States National Academy of Sciences; similar tables to these for wheat, maize, barley, millets, sorghum and grain legumes are also in the report (NAS 1978). The figures in the table are sometimes based on actual measurement (eg the Indian ones) but more usually on expert opinion. They are sometimes specific to one operation (usually storage) and sometimes not specified by operation at all (most of the African figures). One third of the operation specific estimates are low (1-3 per cent) and suggest that intervention costs would have to be very carefully appraised before deciding loss prevention is a priority use of scarce development resources. In practice however, the higher estimates have generally been taken as evidence in defence of funding for PHLP programmes. Low loss figures have prompted occasional cautionary notes (eg Spensley 1977) to the effect that losses may not always be high and that economic and social factors must be considered in organising loss prevention programmes, but such notes have invariably been parenthetic within an argument for promotion of loss prevention programmes. Most significantly, there has rarely been any

TABLE 2.1

Reported Losses of Rice Within the Postharvest System up to 1978

Region/Country	Total per cent weight loss	Remarks
West Africa	6-24	Drying 1-2; on-farm storage 2-10; parboiling 1-2; milling 2-10 (van Ruiten, 1977)
Sierra Leone	10	
Uganda	11	
Rwanda	9	
Sudan	17	Central Storage
Egypt	2.5	(Kamel, 1977)
Bangladesh	7	
India	6	Unspecified storage
	3-5.5	Traditional on-farm storage (Boxall and Greeley, 1978)
	1.0	Improved on-farm storage (Boxall and Greeley, 1978)
Indonesia	6-17	Drying 2; storage 2-5
Malaysia	17-25	Central store 6; threshing 5-13
	c13	Drying 2; on-farm store 5; handling 6 (Yunus, 1977)
Nepal	4-22	On-farm 3-4; on-farm store 15; central store 1-3
Pakistan	7	Unspecified storage 5
	2-6	Unspecified storage 2 (Qayyum, 1977)
	5-10	Unspecified storage 5-10 (Greaves, 1977)
Philippines	9-34	Drying 1-5; unspecified store 2-6; threshing 2-6
	up to 30	Malaysia workshop (FAO, 1977c)
	3-10	Handling (Toquero et al, 1977)
Sri Lanka	13-40	Drying 1-5; central store 6.5; threshing 2-6
	6-18	Drying 1-3; on-farm store 2-6; milling 2-6; parboiling 1-3 (Ramalingam, 1977)
Thailand	8-14	On-farm store 1.5-3.5; central store 1.5-3.5
	12-25	On-farm store 2-15; handling 10 (Dhamcheree, 1977)
Belize	20-30	On-farm storage (Cal, 1977)
Bolivia	16	On-farm 2; drying 5; unspecified store 7
Brazil	1-30	Unspecified store 1-30
Dominican Republic	6.5	On-farm store 3; central store 0.3

Source: NAS (1978) p.65. Based on FAO (1977) figures unless otherwise indicated.

attempt to review evidence on the costs of loss prevention or on the distribution of gains (and losses) between the hungry and others.

2.4 ATTITUDE-FORMING INFLUENCES ON THE DEBATE ON LOSSES

The postharvest lobby was strengthened by the relative invisibility of postharvest losses. This invisibility commonly resulted in neither Ministries of Agriculture, responsible for food production, nor Ministries of Food, responsible for food distribution, having had responsibility for losses in the postharvest system. Realisation of this omission was more persuasively indicated by dramatic statements of high (though unsubstantiated) estimates of losses than by cautious admonitions to improve things by making accurate estimates of loss. As in the Indian case, writers who reported losses were low and did not merit loss prevention programmes were often simply not heard. Two scientists of repute had stated quite clearly that food losses in traditional peasant farming were rather lower than most other estimates. Maize was analysed by Miracle (1966, p.243) and his conclusions from a study of eight countries in tropical Africa was that storage losses - which had been considered by some to be the most important source of loss - were generally negligible. Grist (1975, pp.401-2, but first published in 1959), one of the world's acknowledged experts on rice, estimated farm-level storage losses for that crop in the South Asian context to be below five per cent and that, for stocks held by cultivators, 'no question of more ambitious methods of storage is presented'. These figures had both been available in the mid-sixties. Yet, partly due to the non-involvement of their authors in the postharvest scene and partly because of when and where they were published, they became a 'neglected dimension' in the debate on losses.

Four other factors were important in making postharvest loss prevention fashionable: technological determinism; the impact of the green revolution; the pressure from food donors; and, commercial pressure.

Technological Determinism

Traditional postharvest practices were regarded as synonymous with inefficient practices and it was believed that modernising technical change in postharvest practices offered a straightforward answer to the problems of world hunger. There were at least three reasons why such technological determinism was rife in the early postharvest studies. First, from about 1965 onwards there had been the evidence of the impact of the new seeds; whatever their social and economic impact ultimately, there is little doubt that, in their absence, the numbers of hungry people would have been far larger. This established, at least at a popular level, support for the view that the level of technology was important rather than the circumstances of its introduction to particular societies. The proponents of postharvest technical change were anticipating rapid and clearly visible improvement in food availability as a consequence of such change. The detailed social and economic research on the distributional implications of the green revolution - both by region and by economic class - and on the longer term implications of green revolution technology did not seriously influence the postharvest debate on neglect. The processes by which, for example, drought resistant varieties became a breeding priority did not influence the structure of postharvest research noticeably; ie questions of income distribution effects and of "appropriateness" have not engaged the attention of most postharvest specialists who are seeking technological solutions to problems which they, as agricultural engineers and food technologists, regard as purely technical.

Secondly, the interdisciplinary basis required to effectively analyse loss prevention was not encouraged by institutional structures of research. Instead, single discipline rather than multiple discipline, and laboratory rather than field research were encouraged by these structures. The career structures of individuals and the professional goals of their disciplines encouraged specialism rather than open-ended interdisciplinary research. The consequence was that technical solutions were developed without regard to their costs or to the problems of their diffusion.8/

Thirdly, in common with other areas of rural technology the particular structure of innovation was closely influenced by the structure of dependence on aid. The impact of aid, for example, upon the rate and direction of tractorisation in agriculture has been fairly well documented (eg Burch 1978) and similar events occurred with postharvest technology.9/ The foreign exchange costs of a silo appeared an attractive benefit from aid when the alternative was no aid; yet aid funds to research local innovation were scarce. In addition to problems associated with tied aid, dependence on assistance also affected the professional values of third world researchers. Individuals, equipped with degrees in agricultural engineering from western universities and anxious to establish themselves professionally, rightly perceived that this would be more quickly achieved by application of their sophisticated knowledge of modern technology rather than through development of local innovation. Frequently, the technical research was undertaken by expatriates or even undertaken in developed countries. Inventions in Canberra, Culham and Kansas were to be the loss reducing innovations for the villages of Africa and Asia.10/ A careful reading of the evidence, such as that in Table 2.1, would suggest that situation specific evaluation of alternatives may be rewarding, but funding for that type of research was the exception rather than the rule.

The Impact Of The Green Revolution

The introduction of new seeds had two very important consequences for postharvest practices. First, the increases in land productivity led to more than proportionate increases in the marketed surplus. As Gill (1982,p.154) has shown for one Indian State, a 100 per cent increase in land productivity over four years can result in a 400 per cent increase in marketed surplus.11/ The inability of the marketing system, both public and private, to handle the increased quantities of food grains resulted in, sometimes very severe, loss. Whilst most of the problems caused by increased volumes of production were actually felt at the market place, one of the consequences was an increased concern with the problem of farm-level food loss.12/ More effective use of farm-level storage and handling facilities could reduce the concentration of market arrivals in the public and private sector marketing system but would be possible only if farm-level storage was not associated with substantial food losses.

Secondly, the new seeds were directly a cause of increased risk of loss because they were more susceptible to pest, particularly insect, attack. This increased vulnerability was a natural result of breeding for high ratios of edible to dry matter. Rice varieties often had thinner husks and maize and wheat varieties were often softer;13/ this was particularly a problem for maize but there were also rice varieties that farmers would not store because of their greater susceptibility to insect attack.14/ The emphasis in postharvest R & D was on rice, wheat and maize anyway because of their widespread importance as dietary staples, even though it was recognised (eg NAS 1978, p.2) that the perishables almost certainly suffered significantly higher levels of food loss; the introduction of HYVs reinforced the claim of these cereal staples for special attention.

Also, the seasonality of production was often affected by the introduction of new seeds, in particular when it was accompanied by irrigation. In many parts of Asia high yielding varieties are grown in a second season under irrigated conditions in the dry winter season and are harvested during the early rains. These changes in harvest date sometimes drew attention very dramatically to the problems of postharvest management. In Bangladesh, for example, the Ministry of Food was ordered by the President to purchase wet HYV grain of the boro harvest from farmers in 1979 even though the condition of the grain contravened their usual quality regulations; the reason being that the Government had guaranteed to purchase grain from the farmer.15/ Partly in response to these changed conditions of production, there were few agricultural engineering departments in the Universities of Asia which were not attempting to develop a system of grain (particularly rice) drying, often with support from nationally or regionally coordinated research networks.16/ Whilst, for rice at least, there was little evidence of breeders responding to these problems by altering their screening procedures to include postharvest characteristics, there is no doubt that the green revolution made an important contribution to the advancement of postharvest R & D activity.

The Pressure From Food Donors

One of many aspects of the politics of food aid is the concern of donor countries at the supposed wastage by recipients of food aid. Reports in the American press that food aid given to India and Bangladesh was being eaten by rats did not encourage American public opinion in favour of food aid; to the contrary, it provoked a US Senate Foreign Relations sub-committee enquiry into the incidence of loss by wastage of food aid. The report (US Government 1976) was followed in 1977, by a letter from Senator Humphrey to the Administrator of AID reporting the views of the Committee on Foreign Relations regarding the International Development and Food Assistance Act; it singled out postharvest food loss reduction as a special area of concern - one of only three so treated (Humphrey 1977).

Whilst concern with loss of food aid was not directly related to the farm-level it nevertheless promoted concern with the farm-level postharvest system. It was recognised (eg FAO 1975) that public sector systems might frequently be suffering high losses relative to farm-level because of equipment, management and capacity constraints; however, with the majority of production remaining at farm-level in traditional stores and with the long-term reduction in food-aid as a target, farm-level losses appeared to be a priority concern.

Commercial Pressure

The introduction of new methods of postharvest processing is one of the commonest features of the early stages of rural transformation. In agricultural communities increased crop production is a condition of economic growth and this of course involves a commensurate growth in crop processing. Historically, the development of custom operated crop processing facilities was one of the most important avenues for rural entrepreneurship. The use of the water wheel for grinding corn in the 1st century BC in Greece (Marx 1970, p.408), the application of steam power to threshing drums in 18th century England (Ward 1982, p.45) and the use of diesel-powered rice mills in 20th century Bangladesh were some of the earliest applications of mechanical power in food production in those economies. Entrepreneurs obtain lucrative investment opportunities through demand for new dryers, threshers, improved stores and improved milling which have usually borne a strong relationship to growth in crop output - especially as commercial rather than subsistence farm production occurs. Rice milling in Asia is an important example of this process.17/ Of course, the development of commercial involvement is dependent upon profitable investment opportunities, both for the manufacturer of the equipment and the final user, but there is substantial evidence, with choice of postharvest techniques as with other techniques, that commercial pressure results in technical choices that are sub-optimal (Stewart 1978, p 202). An important sales strategy for manufacturers has been to exploit the view that food losses are high in traditional systems.18/ This has been particularly effective when selling to public sector agencies responsible for crop processing of procured grain. Also, the banks who have financed local entrepreneurs have often regarded the supposed loss-preventing features of modern postharvest equipment as a critical factor in making decisions about financing such investment.19/

To summarise, the development of PHLP programmes occurred in the 1970s in response to a popular belief that the postharvest system was a cause of substantial food losses. Growing alarm concerning projected food deficits and high estimates of postharvest food loss prompted the rapid growth of intervention programmes. The actual evidence on food losses was generally of poor quality, at best unreliable as an indicator of possible programme benefits and also far more mixed regarding the levels of losses than many commentators on loss prevention potential suggested. Genuine, but exceptional, examples of high loss - such as the upper bounds of the Philippines figures discussed earlier - had a disproportionate influence in determining the direction of R & D initiatives towards technical research designed to prevent loss rather than towards analysis of the size of loss and the costs and benefits of loss prevention. Technological determinism, the HYV seeds, aid dependence and commercial interests all contributed towards this research bias.

2.5 THE NEGLECTS OF THE DEBATE

The previous three sections examined the reasons why increased support for postharvest research and development activity was centred on the belief that modernising technical change to reduce food loss should be the principal focus of planned intervention in the postharvest system; this section and that following look at some of the inadequacies of this belief. The central inadequacy was the frequent failure to analyse the cost effectiveness of proposed interventions. Commonly, neither the economic consequences for specific occupational groups notably farmers, village artisans and postharvest labourers, nor the social profitability compared to alternative use of scarce development resources were seriously analysed.

An increase in food availability is only one possible consequence of farm level postharvest technical change; there is no *a priori* reason for assuming that it will be a determining consideration in an investment decision. At the farm level, such decisions depend upon the value of increased food availability compared to the labour and capital costs incurred in achieving it. In most situations it is extremely unlikely, because of diminishing returns, that the optimal farm level investment decision will be consistent with a decision based on minimising food losses. At the social level, both efficiency and equity considerations introduce trade-offs between the food availability objective and broader development objectives. Loss-reducing technical change in postharvest systems may not be the most efficient route to improving food availability and even where it is, such change may have costs in terms of income distribution consequences that result in a social benefit-cost ratio below one.

The analytical tools to address these issues were available; the economic literature on choice of techniques includes many analyses of developing country rural technology choices. In particular, the evaluation of farm level investment choices relating to mechanisation of production has produced extensive empirical evidence on factors affecting the choice of techniques. Farm size differences in access to labour, to capital and to specific farm inputs, economies of scale and risk are important amongst these factors; they have been extensively analysed in evaluating tillage, irrigation and fertiliser techniques and more generally in cropping and farming systems research; but farm-level postharvest techniques have rarely been seriously evaluated in these ways. Improved crop threshing, drying, storage and milling techniques were often advocated and in some cases actively promoted without any economic analysis either of the farm level investment decision or of the social costs and benefits.

The main exceptions to this pattern were studies of storage and of milling technologies concerned principally with public-sector postharvest investment decisions that partially affected the operation of farm-level postharvest systems. These effects related usually, though not always, to the assumptions made about the handling of marketed surplus. On storage, there were studies undertaken in Nigeria (Upton 1968) and for developing countries generally (IBRD, 1970) on choice of public sector storage technology where the need to remove the surplus from the farm level was presumed necessary and was a factor affecting the scale of technology selected for the public sector investment. Important exceptions on choice of milling technique came from Bangladesh, India, Indonesia, Kenya and Sierra Leone20/ in the period 1974 to 1978. With the exception of the Kenyan study where maize was considered, the crop concerned was rice. With the exception of the Bangladesh study, the principal focus was on the net social effects of technical choices rather than farm-level decisions. All the studies, in addition to an analysis of capital and labour inputs, also took account of differences in milling outturn (ie relative food loss characteristics). They were all undertaken by expatriate academics and, with the exception of the Sierra Leone study, were ex post analyses of a pattern of technical change already well developed. There is no evidence that the banks, food corporations or food ministries, that were responsible locally for financial analysis of milling technology investment decisions, utilised these studies even though their results suggested a change away from modern, capital-intensive, large-scale processing facilities to intermediate and in Sierra Leone, hand-processing technologies. The advocacy of large-scale modern units was heavily influenced by the opinion of milling experts who cited the higher milling output (the rendement rate) as a justification for them with scant reference to costs, benefits or distributional impact.

It would be easy but probably wrong to presume that a conspiracy theory could explain the existence of this technological bias. The emphasis upon loss-reducing technical change for developing country agriculture was not ideological in the sense that it represented a class interest but was symptomatic of the developed country origin of recommended alternatives and of the disciplinary narrowness that largely restricted research on postharvest technology to agricultural engineers, entomologists, cereal chemists and cereal technologists. Above all, it resulted from the failure of the social science disciplines, particularly of the economists, to develop a framework of analysis to assess the relative benefits from food loss prevention in financial, economic or social terms. The next section attempts to outline a framework of this sort through discussion of the benefits and costs of postharvest technical change. The detailed specification of a framework cannot be undertaken independently of analysis of specific situations; such a framework is developed for Bangladesh in the next chapter and here the broad features of such a framework are outlined.

2.6 THE COST AND BENEFITS OF LOSS-REDUCING TECHNICAL CHANGE

Investment in postharvest activities is essentially different to investment in agricultural production because it is not possible to increase the volume of output above that entering the system. The gross value of output is fixed and the net value of output differs from it by an amount equal to the postharvest costs plus value of losses (physical loss and qualitative deterioration). In the absence of food losses the best techniques are the cheapest techniques per unit throughout. Where food losses occur they are worth preventing only to the extent that the value of losses prevented is greater than the additional costs associated with the loss-preventing change. This seemingly straightforward framework has a number of complicating elements both for the individual adoption decision and in analysing the net social worth of postharvest technical change. These are examined in turn but first there are three considerations common to both situations that have been and continue to be a source of difficulty in identifying cost-effective interventions.

First, and most fundamentally, uncertainty or ignorance of the levels of food loss in a traditional postharvest practice will prevent 'net food availability' comparisons being made even when the losses likely to be associated with the new practice can be measured. This constraint has been the single most important drawback of recommendations relating to farm level postharvest technology. Secondly, in some operations losses under one practice can be uncertain because they can vary dramatically by season, eg crop drying on large farms, and therefore the most cost-effective practice may be calculable only as a probability. Thirdly, between operations there may be an interaction such that the most cost-effective practice for one operation may be sub-optimal when the cost-effectiveness of the postharvest system is considered; two important examples of interactions like this for rice are the maturity at harvest with drying requirement and drying method with milling yield.

Even when uncertainty and interactions do not complicate the analysis accurate identification of the most cost-effective practice depends upon correct valuation of inputs and outputs. Neither at the level of the individual nor in a rural planning context is this straightforward.

The Farm Level Investment Decision

Conventional neo-classical choice of techniques analysis provides a method for identifying the technique which, at a given level of gross output, enjoys the highest level of value added. The optimal technique is that in which the marginal physical rate of substitution of capital for labour in production is equal to the ratio of capital to labour cost per unit. This is usually presented in a format similar to Figure 1 where the tangency between the iso-cost (equal cost) and iso-product (equal output) curves represent the optimal factor mix in production technology. Real world situations do not generally enjoy a continuous smooth iso-product curve; Timmer (1974) gives an example for rice milling from Indonesia, reproduced as Figure 2, where there are five production possibility points with different capital-labour ratios.

If one technique suffers more food loss than another it will require more grain input to achieve the specified output level. Losses will therefore be accounted for as an additional capital cost. Russell (1980) provides an example of how this could affect selection from amongst alternative drying techniques.

Note also that Figure 2 can be used to indicate 'research bias'21/ - pushing some points on the iso-product curve (e.g. large bulk facilities) inwards towards the origin over time but neglecting others (e.g. the small mill). The data needed to test the strength of this bias for postharvest techniques are unfortunately not available (though for the Bangladesh case the very fact of a move towards less labour intensive techniques when real wage rates have been falling - see Chapters 3 and 6 - is itself evidence of a capital-using bias.)

This neo-classical approach, when applied to developing countries, has focussed attention on employment objectives, savings contraints and price distortions resulting in sub-optimal technical choices. However, as Stewart (1978 pp.25-29) discusses the framework only addresses two characteristics (labour and capital requirements) of a technique whereas there are many other characteristics which affect the choice of techniques; Stewart lists (p.26) scale of output, nature of product, skilled labour requirements, material inputs, infrastructural requirements etc. etc. All of these factors are relevant to the choice of postharvest techniques and their significance is considered in Chapters Five and Six on choice of threshing technique and of milling technique. The first characteristic Stewart lists, scale of output, has been of special importance however for two reasons.

THE SCALE PROBLEM: First, the adoption of any technique, eg in crop drying, where the capacity utilisation is low because grain input is small will necessarily increase the net costs of using it. In effect, this has restricted the farm level choice of techniques, precisely because the range of alternatives has very often been of inappropriate scale. This factor has been responsible for a shift away from farm-owned processing equipment to custom-hire of such equipment. In Bangladesh,

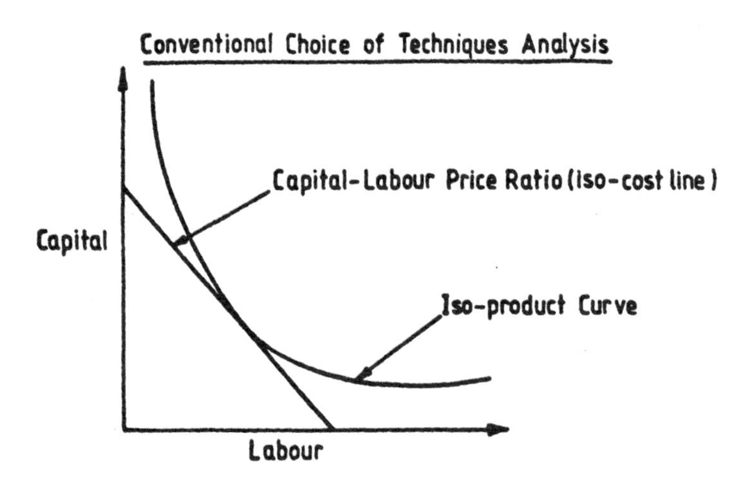

Figure One

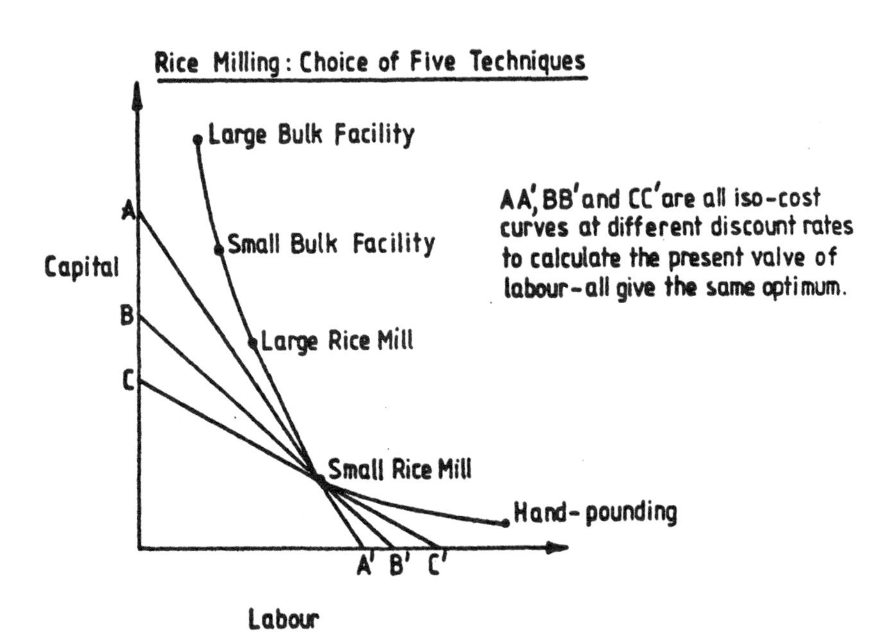

Figure Two (from Timmer, 1974)

custom-hire of threshing, drying, storage and milling equipment has been organised - and in the case of drying and storage promoted by postharvest teams explicitly concerned with food loss reduction. Similar patterns of commercialisation have been a frequent response elsewhere to the scale problem with drying, threshing and milling techniques and constitute a major characteristic of postharvest technical change. The choice of techniques becomes a two stage exercise with both rural entrepreneurs (sometimes large farmers) and farmers making separate calculations of the costs and returns to adoption. Both the examples of technical change examined in this book are of this sort and the employment and income distribution consequences of such structural change are examined in detail below (see Chapters 5 and 6).

A second main response to the scale problem can be seen in the thrust of postharvest R & D towards scaling down of existing techniques which have been profitably operated on larger grain inputs.22/ In Bangladesh, threshing, drying, storage and milling equipment have all also been the focus of such research. As discussed below (Chapter 3) these have not been successful but they have been useful in drawing engineers' attention to the trade-offs between cost-criteria (notably, engineering or theoretical throughput rates) and performance-criteria (notably, actual throughput rates) which are a central issue in planning postharvest technical change and the neglect of which has been a major stumbling block to technology diffusion.

FARM-LEVEL CONSIDERATIONS: The choice of techniques framework focuses analysis on the capital and labour markets and addresses the issue of food loss only indirectly insofar as it is one of the economic criteria affecting farm-level postharvest technical change. Whether commercialisation through custom-hire operations or individual farm purchase of new equipment is being considered the choice of techniques by the farmer will be a response to his overall assessment of economic return as mediated through his relations within these markets. Moreover, farmer choice of technique will also depend upon alternative uses of his productive assets and his investment behaviour will reflect the full range of opportunities open given his economic strategy (eg profit maximising, minimum income satisfying or risk minimising).

These seemingly obvious considerations have not always been acknowledged. When farm level questionnaire surveys have been undertaken they have often been informed by a view that either the farmer is ignorant of how severe his problem of food losses is or that he is unhappily aware of it but ignorant of loss-reduction possibilities, which he would eagerly adopt once made available. And the design of the questionnaire, the structuring of questions and the analysis of results have often been oriented towards confirmation of the interviewer's (or his project leader's) viewpoint. So strong is this bias that there are examples where, when the questionnaire results did establish a relatively low level of perceived losses, the results were ignored; a Malaysian survey (Mohamed 1981) of this sort indicated total loss levels of around 7 per cent across all rice postharvest operations except drying and milling but the author, in discussing recommended interventions, referred only to a foreign expert's guestimate of 25 per cent losses - and until questioned seemed unaware of the explicit bias of his analysis.

Pre-occupation with loss prevention carries the implicit presumption that the neglect of the issue of food loss has been shared by farmers also. This concept of farmer neglect parallels the conception of peasant economy (see Sahlins 1972, Chayanov 1966) as one of utility or needs satisfying where self-sufficiency and a subsistence minimum is the object of farmer endeavour. The evidence presented in subsequent chapters suggests that this is not so and that "risk-averse profit maximising" more accurately describes farm level behaviour in choice of postharvest techniques. Whilst this view represents a departure from conventional wisdom judged by the narrow perception of many commentaries on postharvest problems it conforms to models of behaviour (Lipton 1968) that have long been accepted as a correct interpretation of farm-level investment in the sphere of production. The structure of rural capital, labour and product markets, which determine the farm-level assessment of technical choices, have been analysed in minute detail in relation to innovation in the sphere of production; explanations of superficially irrational farmer behaviour caused by distortion of markets away from theoretical norms of perfect competition are central to the planning of rural development. It is not appropriate here to attempt to summarise or reproduce any part of that literature. Rather, ten common conditions relating to capital, labour and product markets and which characterise postharvest technical change are outlined below to give some indication of the considerations involved in the farm-level investment decision.

a) Traditional postharvest techniques use very low levels of capital and have low levels of labour productivity. This is also true of many production techniques - but when they are modernised the increase in output can outweigh the fall in labour per unit of output and result in an increase in labour use. But postharvest technology doesn't raise output. In the absence of food loss prevention therefore, any cost reducing technical change will necessarily be labour replacing.

b) Technical change that <u>always</u> reduces food loss is inherently risk-reducing. In other words, the outcome of the investment decision is predictable - unlike in farm production where investments, eg HYV seeds, often increase the degree of uncertainty over future income associated with a particular level of outlay. Where losses are themselves highly variable, eg in crop drying, this advantage does not exist.

c) A technical change that reduces postharvest food loss and benefits own consumption will for many farm households in effect extend the period of self-sufficiency; the period of dependence on consumption loans or market purchase will be reduced and thus the benefits of loss reduction will be realised precisely when food is most expensive to obtain. This seasonal effect will most benefit those households where some consumption normally has to be met from non-family farm purchase - and for those now able to store and sell larger quantities at higher lean season prices. If the consumption shortfall was normally met either by eating less food or eating inferior food this argument has to be modified:23/ in such cases the net monetary benefits of the new practice will be less in proportion to the differences in costs associated with the different responses to consumption shortfall. However, from a nutritional perspective the outcome is open and may result in the same, worse or better nutritional status: if the additional food results in meeting calorie deficits it will be a net gain; if the inferior foods - ie foods that are purchased in inverse proportion to income - are in fact more nutritious, as is often the case, then a worsening of nutritional status could occur. Expected improvement in nutritional status as food intakes increase does not always occur because of consumer preferences. People at high nutritional risk are unlikely to be irrational in food matters however and those who switch readily to superior foods are probably receiving sufficient calories or have below average requirements.

d) A related point, in the product market, is that the realisation of economic returns from food loss prevention frequently depends upon the sensitivity of market prices to the quality of grain. Sophisticated urban consumers who require whole highly polished rice may pay very low prices for an undermilled rice with many broken grains but these standards often do not apply in the rural markets, especially not where the grain is for home consumption. Similarly, where machine analysis of moisture content is not undertaken, differences in moisture content of two or three per cent are not easily detected - and even less easily agreed upon.24/

Consumer-observable quality standards have different interpretations; for example some South India farmers pay more for paddy which shows evidence of insect attack; their preference in rice consumption is for grain that is aged and to them the best guarantee of this is evidence of insect attack.25/ Arbitrary standards of quality that are not reflected in market prices will therefore give misleading evaluations of the benefits from 'loss-reducing' technical change.

e) The capacity of a farmer to adopt a specific practice, even when it represents a clear and undisputed increment to the final value of crop output may be constrained by the cash requirements. This is obviously true in cases where absolute poverty restricts access to the investment funds needed but even in situations where the farmer's nominal income in the form of crop production is not a constraint, his access to cash may be for two reasons. First, the transaction costs of converting farm output into cash may be so high that the value of additional output from loss reduction is no greater than these costs; this problem has sometimes been observed in the use of insecticides in grain storage. Some farmers in Tanzania, for example, could apparently reduce insect damage cost effectively by selling a small part of their maize in store, treating the rest and ending up with more maize for consumption than if the small harvest-time sale had not occurred. Whilst they can barter small quantities at village level, to obtain cash they have to sell to traders - legally in fact they are supposed to sell only to the government - who are often far away and who often do not deal in such small quantities as the average small farmer would need to sell to buy his insecticidal dust.26/ Secondly, and a more general problem, the realisation of the benefits in the form of more food consumption is the most likely outcome of loss reducing technical change for small farmers and this may not represent a priority use of his scarce cash resources - loan repayment, production inputs and other consumption items may all have a higher priority.

f) As outlined in point a) above, a common characteristic of postharvest technical change is a reduction in labour demand but where the saving is of family labour the market valuation of this labour saving may not reflect the true opportunity cost of labour to the farmer. Shortage of wage employment opportunities may prevent or social relations may inhibit - and in the case eg of Muslim women in Bangladesh, commonly prohibit- non-family farm employment; if alternative productive employment in the family farm is not then available the opportunity cost of family farm labour may be little above zero. This will be particularly relevant in predominantly female activities such as hand-pounding where the work is not seasonally bound and where gender relations of production may restrict alternative forms of female employment or male perceptions of its value when it is not directly remunerative.

Market valuation of labour can also overstate the true labour cost when permanent or semi-permanent hired labourers are employed; social relations may dictate that the farmer is under an obligation to employ them27/ and will not immediately benefit from a reduction in their labour effort when other work is not available.

Finally, a change in technique may involve a reduction in the total labour demand but a time-specific increase. For example, the operation of pedal threshers is most efficient when two, and ideally four, labourers work together whereas under a traditional system of hand beating or bullock treading two labourers for one day will thresh about as much as one labourer for two days. Such time-specific demands can also affect improved drying and milling techniques and may require the farmer to hire labour for work that would otherwise have been completed by family or permanent hired labour.

g) Similar considerations to those in point f above also commonly affect the use of additional capital where transactions costs and time preference may significantly alter the costs of capital relative to market interest rates; these of course affect all investment decisions and are not specific to postharvest investment. However, a particular feature of postharvest capital assets in traditional postharvest techniques and in some cases in new techniques, is their multiple use. Tractors and bullocks are used for ploughing and transportation as well as threshing; farmyard space is used for threshing as well as drying crops. A farmer's decision not to use his bullocks for a second threshing but to plough in preparation for his next crop may result in increased food loss because of mould development and germination of the unthreshed grain but the net economic return from 'neglecting' this task may be substantially greater if as a consequence he can sow or transplant his new crop under ideal conditions. A failure to follow the ideal postharvest practice is not necessarily due to either neglect or ignorance, and is unlikely to be either of these. It is because the net return on using scarce farm assets elsewhere is greater.

h) Similarly, the farmer himself as business manager and as decision taker may not direct his management concern towards postharvest loss prevention if other more productive opportunities require his time. This may sound implausible or at best insignificant but was surprisingly common at the harvest time in Bangladesh when decisions in the field, and supervision of labour there, were urgent. Recently harvested untreated grain in stacks may deteriorate because of mould development and germination but this may be of less economic consequence than failure to plant the next crop in time. The same point can apply, though perhaps with less force, in farmer decisions on the acquisition of both knowledge and the means to reduce postharvest losses compared to, say, deciding upon and acquiring new varieties.

i) Where the perceived risk of postharvest food loss is high a farmer may decide to market his threshed but unprocessed crop (and buy back consumption needs) rather than invest in loss-reducing technology. This is a frequent occurrence with HYVs which are known to be difficult to store (Boxall et al 1978, p.186); in other cases - though not always to prevent loss - farmers sell the standing crop and thereby rid themselves of all postharvest responsibilities. The economic logic of such practices is not easy to analyse because it involves a comparison of the value added where his management and his postharvest equipment may have been used elsewhere on the farm and their opportunity cost is not simply calculable. It is likely though that this marketing option will be a main consideration in drying and storage investment, particularly for large farmers for whom the decision is not if to sell but when to sell; the trade-off in such cases is between the likely size of the seasonal (and quality) price differentials and the costs of investing sufficiently to be able to take advantage of them.

j) Finally, the perception of food loss will determine the nature of response by the farmer. The promotion of loss reducing technical change by rural banks and extension services will be based on a conception of the dimension of the problem which may very well not be mirrored exactly in the farmer's own view; moreover, both views are unlikely to be typical of actual food losses which are difficult to estimate exactly, which vary according to the amount of effort put into postharvest management by the farmer and which also vary by season and by variety. The variation of loss on individual farms around the mean loss on all farms will, by virtue of the nature and causes of food loss, cover a range which in many cases will include the cross-over point where loss reducing technical change moves from being economically beneficial to being non-viable. Specifically, food losses are not normally distributed around the mean but skewed positively. Thus, the modal value will be below the mean value. In other words, average levels of loss, even when estimated carefully, may be a poor guide to the economic benefits of loss reducing technical change for individual farmers. A related problem is that of non-economic motivation for specific choice of techniques; the status value of using postharvest equipment which is different, and more modern, may influence early adopters more than the apparent economic benefits. This will be particularly true when extension workers, publicly employed or commercial, are encouraging adoption for demonstration purposes. Such conspicuous consumption motivation may also influence farmers where the economic benefit is marginal. The early adoption of metal bins in part of Uttar Pradesh, Haryana and Punjab has probably been influenced by such considerations.28/ In addition, non food-loss characteristics of technical change, in this case compactness, cleanliness and ease of installation, exert an influence of unknown strength upon the adopter's decision.

It will be apparent from these ten considerations that accurate assessment of farm-level postharvest loss prevention intervention prospects is a complex matter. Even seemingly advantageous characteristics such as risk-reduction and benefits accruing in the lean season have been seen to require substantive qualification. Together, they represent a serious challenge to postharvest orthodoxy which, however well motivated, has advocated technical solutions without serious analysis of their economic merit.

The Social Worth Of Farm-Level Technical Change

Technical change to reduce farm level postharvest food losses has to be evaluated for its effects on all the objectives of national development policy to provide a comprehensive analysis of social cost-effectiveness. At the broadest level national development policy can be viewed as having a single objective - an increase in the level of aggregate consumption. The techniques of project analysis provide a planning mechanism to quantitatively identify those activities which best contribute to the desired distribution of consumption amongst current consumers and between current and future consumers. These techniques can be applied to evaluate whether specific postharvest technical changes are of net social worth by measuring their effects on aggregate consumption.

In this section, the relationship between this consumption objective and loss-preventing technical change is considered. The discussion is not concerned with the details of how the objective is made operational through project appraisal methods but with the principles embodied in those methods and how their incorporation in a choice of techniques analysis affects the choice of optimum techniques. The issues involved are not peculiar to postharvest choice of techniques but relate to the more general question of optimal factor use; i.e. how optimal capital-labour ratios are affected by the move from a private profitability to a social profitability analysis.

However, underlying the discussion is the recognition that traditional farm-level postharvest techniques are typically labour-intensive and that new postharvest techniques, whatever their food-loss characteristics, are likely to be labour-saving and capital-using.

The issues raised are relevant both to the appraisal of a proposed public sector project to reduce food loss at farm level and to the evaluation of private sector initiatives in use of new postharvest methods. The evaluation of costs and benefits differs according to whether they affect public or private consumption or public or private investment but the principles underlying such valuations are constant - and reflect the aggregate consumption objective assumed to underline all economic planning. If, as in the Bangladesh case, the introduction of new postharvest techniques is occurring through private sector initiatives rather than in the form of government planned intervention, the government can use market controls (pricing, licensing, quotas, taxation, etc) to encourage or discourage particular techniques. To the extent that such controls can be made to operate effectively it is possible to intervene and direct private investment resources in an optimal allocation, though clearly, the probable effectiveness of such controls has to be realistically assessed and associated additional costs must be included in the social cost-effectiveness analysis. The following discussion does not make any assumptions about the source of investment.

How might the aggregate consumption objective be affected by a project to reduce farm-level postharvest food losses? To answer this question the most useful point of departure is to consider the private profitability to an investor of a particular project and then to determine what changes in the valuation of costs and benefits are necessary in order to measure social profitability. There are three types of adjustment necessary in the valuation of costs and benefits: adjustments for the distortion of relative prices, for the sub-optimality of savings and for the inequality of current consumption. The application of these three adjustments in project appraisal is dealt with at length in the many manuals on social benefit-cost analysis29/ now available; here, we briefly summarise in turn the principles underlying each of them as espoused by one of the best known manuals (Little and Mirrlees 1974).

PRICE DISTORTIONS: In standard neoclassical economic theory prices are determined by supply and demand and, in equilibrium, relative prices should reflect the relative scarcity of resources - in the jargon of neoclassical economics, price equals marginal social cost. When a price system meets this condition the objective function, which is defined as aggregate consumption, is maximised. There are a number of reasons why market prices, especially in developing countries, may not reflect true marginal social cost of resources such as labour, building materials, machinery etc. The existence of tariffs and taxes and the fact that, generally, land, labour and capital markets in developing countries are imperfect are two of the most well-known reasons for price distortion; Little and Mirrlees (1974 pp.29-36) provide a more extensive list. Therefore, the first adjustment made to transform a private profitability calculation into an estimate of social profitability is to remove price distortions by converting domestic prices into social accounting prices. For traded goods, the accounting price is the international or border price as it is known; for non-traded goods (e.g. transportation or electricity) the price is based on an estimate of its marginal social cost by revaluing all the required inputs at border prices; and, for labour the price is its opportunity cost as measured by the marginal product of labour.

It is unwise to make generalisations about the consequences of adjusting for price distortions on the choice of techniques for a particular productive purpose; however, in labour surplus economies such as Bangladesh these adjustments often tend to favour labour intensive techniques because market prices tend to undervalue capital and over value labour (Little and Mirrlees 1974, Chapter XII and UNIDO 1972, Chapter 4 and 5). This first adjustment is therefore likely to work against the introduction of new farm-level postharvest techniques.

SUB-OPTIMAL SAVINGS: The second adjustment made to private profitability calculations in order to estimate social profitability is to allow for any sub-optimality of savings. The income generated by a project can be used for consumption immediately or can be saved, and then be available for investment which will generate future consumption (and savings). The relative value of present and future consumption depends on the consumption rate of interest30/ (CRI) which is the rate at which the value of a unit of consumption in the future is discounted compared to its value today. The distribution of income between current consumption and savings is dependent on all the individual decisions made by income earners; it is entirely possible, and very likely in developing countries where savings rates are often low (Little and Mirrlees 1974 pp.48-52), that the implied CRI of income earners' collective savings decisions is greater than the government view of the CRI. In such a situation income use decisions are socially inefficient. Savings are sub-optimal because too much income is allocated to current consumption thereby leaving insufficient income available in the form of savings to generate future consumption through investment. (There are theoretical grounds for disputing both the existence (Wellisz 1977) and measurability (Sen 1984 Part II) of the sub-optimality of savings but in this study the usual SBCA method is adopted - see Chapter Six).

If future consumption, and therefore investment to generate it, is valued low relative to current consumption, then - in economies, like Bangladesh, where there is much labour and little capital - a planner would normally select a labour intensive technique. In the extreme case of unlimited surplus labour the maximum output would be obtained by maximising the output to capital ratio; with a zero social opportunity cost (shadow wage) of labour, capital would be combined with labour until social marginal productivity of labour was also zero (see Sen 1975, Chapter 11). More generally, the labour intensity will increase to the point where the social marginal product of labour equals the wage. But, to the extent that the generation of savings that will produce future consumption is inadequate there is a conflict between present and future consumption. The conflict arises because the technique that employs most labour and thereby generates most current consumption (because labourers consume more than capitalists) is not the technique that generates most savings. Of course, the technique that generates the most savings is not desirable either unless no importance is attached to present consumption. If the government were able to use fiscal policy to tax present consumption and thereby make more funds available for investment it would not have to use project appraisal techniques to alter allocation decisions. In practice many developing countries' fiscal policies are not sufficiently effective to do this. Social analysis of project choice therefore has to incorporate a decision on the relative importance of present and future consumption, and typically, this involves putting a premium on savings (future consumption) because the privately determined share of savings in current income is considered to be too low. This premium will influence the mix between capital and labour in the optimum technique because of labour's higher propensity to consume now its share of output.

The application of a savings premium in project appraisal involves revaluing present consumption in terms of uncommitted social income. The social value of wages, which are taken to represent the main part of present consumption,31/ is no longer equivalent to the social marginal value product of labour conventionally defined but is adjusted downwards to express the lower value of consumption compared to savings (uncommitted social income).32/ In other words, by taking account of the rate of savings in planning technical choices there is necessarily a bias away from labour towards capital-intensive techniques. It implies that the social need for increasing savings rates will determine a technical choice with factor proportions more capital intensive than mere adjustment for price distortions would indicate. Whether in practice it would result, in adoption of a more capital intensive technique in, say, threshing would depend upon technological possibilities; with completely variable factor proportions it would clearly have this effect.

CONSUMPTION INEQUALITY: The third adjustment to private profitability is to take account of the inequality of current consumption. In its effect on capital-labour ratios this adjustment exerts a directly opposite effect to the adjustment for savings sub-optimality. There are two ways in which equality can be explicitly included. First, the shadow wage rate can be reduced to encourage more labour intensive projects. This is in direct conflict with the savings (future consumption) adjustment which requires a higher rather than lower shadow wage. Secondly, the social value of extra consumption to different economic classes can be taken account of by giving more weight to the extra consumption of the poor. This adjustment is made by setting a par level of consumption that is equal to the value of uncommitted social income (see footnote 32) and giving all consumption levels below it a higher weight and all above it a lower weight.

The choice of the weight (approximating the elasticity with respect to consumption of the marginal social utility of consumption) is a critical determinant in evaluating the sort of labour-displacing technical choices that are often associated with food loss reduction (see UNIDO 1978, Ch VII). Use of the weight is based on the view that choice of a technique that increases net present value by reducing food losses but at the same time distributes more of that income to rich farmers and less to landless labourers incurs equity effects that most people would consider undesirable.

Several project analysts (eg Gittinger 1972) develop project analysis techniques that explicitly exclude current income distribution effects but include a social opportunity rather than actual cost for capital. The apparent justification for such an approach is that the government is able to distribute current consumption equitably but is unable to influence the mix between saving and consumption. This is unlikely. It ignores one of the only effective means to incorporate an equity concern, through income weighting in project analysis. In the case of the landless poor their loss of postharvest earnings may not be substitutable and income weighting provides the only means to register that by giving a large value to even small changes in their income.

To summarise, social analyses of technical change will, on efficiency grounds, encourage adoption of more capital-intensive techniques than would take place in the market because the planner views market savings as sub-optimal. The effect of this is mitigated in two ways: by the use of accounting prices which usually result in correction for the relative underpricing of capital compared to labour at market prices; and by equity considerations which give more weight to incremental income for the poor and, through a lower shadow wage, can more generally favour labour intensity. Traditional postharvest techniques tend to be labour-intensive; while some loss reducing change can be labour-intensive, eg in storage and drying, most planned postharvest change results in reductions in use of labour. Where labour is in short supply and where inequality in the distribution of income is, or is considered to be, low, loss reducing technical change will not be in conflict with the equity objective - provided of course that it is genuinely cost-effective on the efficiency criteria.

But in many countries, including Bangladesh, equity considerations conflict with capital-intensive technical change and even substantial food loss reductions may not be justified if the consumers who benefit are earning income much in excess of the par consumption levels. The changed distribution of income is therefore central to any evaluation of food loss reduction in such countries.

Finally, it needs emphasising that social benefit-cost analysis, sensitive as it can be to equity concerns, remains an unwieldy and overly aggregative planning tool. The social costs of lost employment opportunities for some occupational groups may be insufficient to prevent a technical change being socially profitable but for the groups themselves the cost may be their whole livelihood -

basket weavers and rice processing labourers are two postharvest groups potentially of this sort. Research on loss-reducing technical change therefore has to be sensitive not merely to the aggregate cost-effectiveness of particular changes but also to the additional programmes needed to secure the welfare of those put under economic duress by such changes.

Employment And Food Self-Sufficiency As Merit Wants

Increasing consumption is the principal purpose of economic activity but sometimes other objectives are regarded as being of intrinsic worth independent of the consumption objective. These are known as merit wants. Employment is sometimes considered to be a goal of government policy independent of consumption though clearly they are very closely related since it is employment which generates consumption. A second merit want, economic independence, is frequently presented as an objective wholly separate to the consumption objective. Economic independence, though a difficult concept to be precise about, is mentioned most commonly in particular sectors where dependence upon foreign supplies represents or potentially represents a threat to the freedom of national economic management. Energy, capital goods and, in the poorest developing countries, food are the sectors where this threat is most obvious. An objective of food self-sufficiency is of particular relevance to farm-level food loss reduction. In this section we consider some implications of employment and food self-sufficiency as merit wants in relation to the conventional project appraisal framework and farm-level postharvest technical change.

EMPLOYMENT: Employment can be explicitly incorporated into the framework of social benefit-cost analysis by the use of an employment premium (Little and Mirrlees 1974 Ch. XIV). This premium could be incorporated in project appraisal techniques when the creation of a job reduces consumption handouts by the government to the unemployed but it can also be used, and this is the case we are interested in here, as a method of increasing employment. In project appraisal techniques a premium of this sort would reduce the social cost of employment by lowering the shadow wage. Use of this premium would presume a value attached to employment wholly independent of any consumption benefits.

The justification for a value of this sort would have to be based on putative advantages of employment such as greater political stability, aiding backward regions or social groups where unemployment is particularly severe or reducing the psychological poverty (lack of independence, sense of non-achievement) caused by unemployment - this last advantage is called by Sen (1975, pp5, 39-40, 81-83) the 'recognition aspect' of employment.

Since the traditional postharvest system at farm-level is highly labour intensive and since, as we have discussed, proposed loss-reducing technical change usually reduces labour requirements, the incorporation of an employment objective would favour continuance with existing techniques. However, there are two reasons for believing that an employment premium should not be used in countries like Bangladesh to influence technical choices such as these and indeed Bangladesh does not use them. First, in the very poorest countries where both consumption levels and investible resources are very low, any policy choices that impinge on increasing aggregate consumption by imposing conditions on that policy objective are unaffordable luxuries; in other words the recognition aspect of employment is something poor countries cannot afford to consider given the overriding importance of the aggregate consumption objective.

Secondly, as Sen describes (1975 Ch.9), it is by no means clear that job creation in specific projects will actually result in increasing aggregate employment. Suppose that use of an employment premium results in a different choice of technique to what would have been the choice without the premium. In other words, but for the premium, this choice would result in a lowering of social profitability either by using more resources for the same output or a lowering of output with the same resources. In both cases use of the premium will ultimately reduce the total availability of consumption goods33/ producible with the resources available to the economy. The extra jobs will create an extra demand for consumption goods but if the constraint on output is not ineffective demand but "resources and organisation" (Sen 1975, p.83) this extra demand will only result in inflationary pressure. Sen uses this argument to suggest that the total volume of employment is determined by the ratio of the supply of wage goods to the real wage rate; with a given real wage rate any policy, such as an employment premium, which reduces total output must also reduce total employment.

Neither of these arguments would apply when the jobs being created are performed by family labour without a wage but they both apply to the use of hired wage labour and it is the displacement of this hired labour which is the most sensitive issue regarding the employment effects of postharvest choice of techniques. Both the reasons advanced for not using an employment premium are based on the view that ultimately more jobs will be created by eschewing use of the premium. Whilst accepting this view in general there are circumstances in which the argument needs some modification. These circumstances are those, referred to in the previous section, where loss of a job results in loss of livelihood.

It is argued in Chapter Seven that displacement of female wage labour in Bangladesh resulting from postharvest mechanisation is a substantial threat to the livelihood of these women - the gender division of labour restricts their access to other wage work, they have limited opportunities for non-wage productive work and their households require a contribution from them to maintain consumption. Yet, the macro-economic nature of the arguments against the employment premium are insensitive to these circumstances which create a level of deprivation and misery no planner could willingly promote. The modification required is the recognition that employment of specific occupational groups may be a merit want. Use of an employment premium in project appraisal techniques will not satisfy this need and instead the cost of providing new work opportunities or otherwise subsidising those displaced should be included as a cost of the new technique that results in this type of labour displacement. This, of course, is not a new suggestion (Sen 1972, p.17).

FOOD SELF-SUFFICIENCY: If food self-sufficiency is a merit want which has absolute priority then the social analysis of choice of techniques becomes a cost-effectiveness exercise where the cheapest method to attain self-sufficiency is adopted regardless of the evidence of more (socially) lucrative investment opportunities in other sectors. Whilst this interpretation of the food self-sufficiency objective has undoubtedly moved postharvest planners, the absoluteness of this priority has to be subject to conditions imposed by the underlying consumption objective. Even if it were not, farm-level food-loss reducing technical change would not necessarily be desirable; improvement in food production may be a cheaper method of reaching self-sufficiency.

Whilst food self-sufficiency, or the less stringent formulation of the objective as 'food security', is generally discussed in terms of food availability per head the underlying issue is food consumption particularly of course the consumption of the poorest. From a food-consumption perspective the food distribution system provides an alternative investment channel to meet the objective. Therefore, both investment in food production and in food distribution may offer more cost-effective routes to food self-sufficiency. Farm level food loss prevention may nevertheless be adopted for two reasons; complementarity with the distribution or production strategies to achieve food self-sufficiency as illustrated in Figure 3 and institutional or supply constraints on adoption of those strategies.

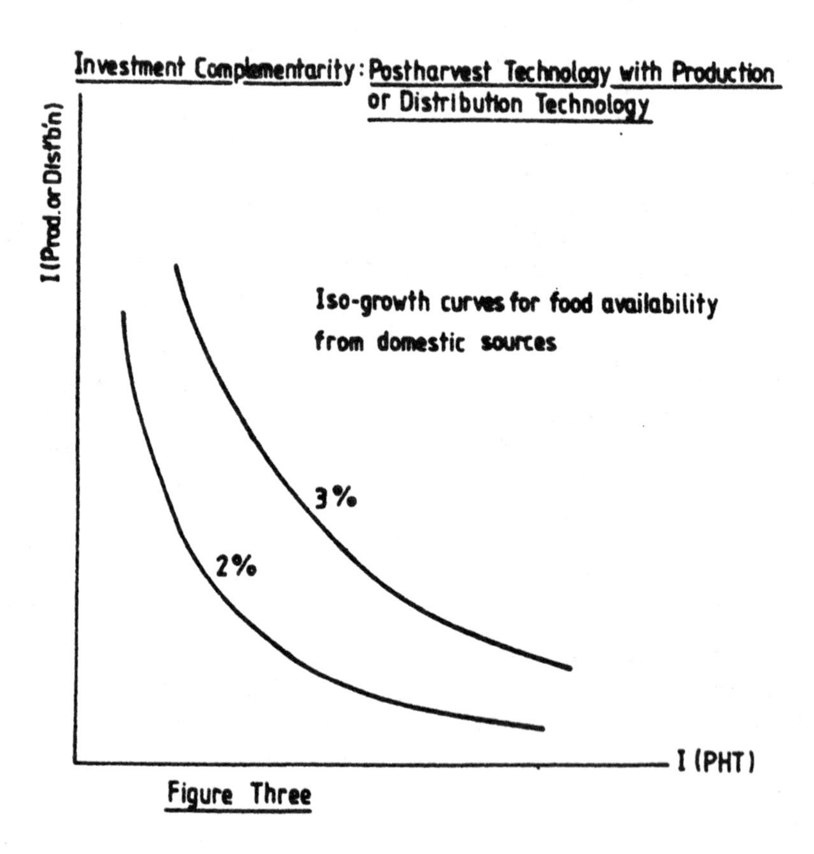

Figure Three

Investment on the distribution side may often be complementary to loss reduction in the marketing and non-farm postharvest system; ie more effective use of non-farm storage, processing and transportation facilities may involve food loss reduction. It is less likely to have any direct effect on farm level postharvest operations, at least in the short-run, unless procurement and pricing policies encourage farm-level adoption of loss-reducing techniques. For example, the introduction of grading standards based on moisture content and head rice yield may encourage use of better threshing, drying and storage techniques. But the direct effect on grain stored and processed at farm-level for seed and own consumption will probably be small until the distribution system is sufficiently reliable and cost-effective to compete with farm-level postharvest operations for provision of on-farm consumption needs. For example, if the price of maize flour in the market is sufficiently low, a farmer may find that it is more profitable to sell his unshelled maize and buy corn flour than to process his own crop; the greater his food losses the more quickly he will make this response to improvements in the distribution system.

Investment in improved production is more likely to be complementary to loss-reducing technical change in the farm level postharvest system. The risk of food losses, as discussed earlier in this chapter, has increased with yield increases because of changed varietal characteristics and seasonality and because the increased throughput may reduce the unit efficiency which can be obtained from traditional post harvest operations. This latter may not imply a change in technique for the farmer but simply more investment in traditional postharvest equipment; more interestingly, it may induce a change in technique that reduces the costs to the farmer through labour savings but actually increases the level of food losses - provided the net effect is greater value added the technique will be preferred. As the Bangladesh evidence presented in later chapters demonstrates, this type of change is occurring; but in the long run, as land productivity grows and postharvest scale economies can be realised, it is inevitable that the level of losses will become a more central component of cost-effectiveness calculations. The desirability of any loss reduction will of course still be conditioned by the effects on the consumption objective.

Institutional and supply constraints may restrict the adoption of the most cost-effective strategy to realise a food self-sufficiency objective. An obvious example from smallholder agriculture is where access to improved seeds and fertiliser is constrained by the performance of credit institutions. Under these conditions, farm-level loss-reducing technical change may be undertaken even though it is not the most socially cost-effective method of reaching the food availability objective.

Thus, there are three circumstances in which a food self-sufficiency objective may determine the promotion of farm level food loss reduction in postharvest operations. Two of these have been discussed; when other methods, cost-effective in isolation, are in fact complementary to postharvest investment so that their adoption induces adoption of loss reduction activities; and when these other methods face constraints on their adoption. The third circumstance is when farm-level food loss reduction is a more cost-effective method than the alternatives.

It is worth noting that in all cases the increase in food availability achieved through farm-level loss reduction may not have an exact correspondence with an increase of food availability that the self-sufficiency objective is directed towards. Most postharvest commentaries on the benefits of loss reduction calculate the value of grain saved at its marginal import cost on the assumption that an exactly equivalent volume of imports will be saved. Now, whilst social cost effectiveness analyses would also use that value - with an adjustment for transport costs and weighted by the value of increased consumption measured in terms of uncommitted social income (see footnote 32) - the justification for doing so is that this represents the social value of benefits, not that a foreign exchange saving has actually been made. It may happen that the farm family which reduces its grain losses consumes exactly that much more without any change in anything else. Alternatively it may substitute consumption of other foods by the newly saved grain. In neither of these cases will there be any direct effect on the amount of grain supplied to or purchased from the market. The precise effects will depend upon the grain saver's marginal propensity to consume grain and the substitutability with other foods. These examples are extreme cases, though in the case of Bangladesh they may often be realistic; the point is that not every increase in food availability will be translatable into a short-run improvement in food self-sufficiency measured by reduced import dependence. In the long run such an approximation has more validity though it will never be exact because of income and substitution effects if there is a price response to the domestic supply increase.

The food self-sufficiency objective does not introduce any changes in the method of analysing alternative technical choices. The difference lies only in the criteria for acceptance; a given food self-sufficiency objective will involve adoption of the most cost-effective methods until the goal is achieved whereas without that objective even the most cost-effective method of improving food self-sufficiency would not be adopted if it fell below the criterion established for project selection. In practice, many countries, including Bangladesh, have a food self-sufficiency objective but are either not able or willing to commit sufficient resources to achieve it, or make optimistic assumptions about the productive capacity of resources so committed. In those circumstances actual behaviour rather than stated objectives may provide a better guide to the governments' objectives; the fundamental principle of appraisal is the same whether food self-sufficiency is the ultimate arbiter of project selection or not.

NOTES

1. This Conference, together with the FAO biennial conferences of 1975 and 1977, led to the organisation of the special FAO prevention of food loss programme in 1977. Other bi-lateral and multi-lateral donor agencies have also singled out postharvest farm level food loss prevention as an area of special interest; UNEP (1983) lists in an appendix many of the international and national organisations involved in such activities.

2. Eight organizations have formed the Group for Assistance on Systems relating to Grain after Harvest (GASGA) which provides a coordinating role (see UNEP 1983 p.61).

3. AID also had Congressional approval for loans of $44 million to Egypt and $60 million to India for grain storage activities.

4. In India, for example, the food production indices for 1972, 1973 and 1974 were below the 1969-1971 index and in 1974 fell as low as 91 per cent of that index. Nigeria, with a 1969/71 base of 100 had food production indices in 1972, 1973 and 1974 of 94, 86 and 91. In Bangladesh the corresponding figures were 86, 95 and 90, 1974 being the year of a major famine. The figures are from the 1980 FAO production yearbook.

5. The Indian Planning Commissions had drawn attention, in reviewing the Fourth Plan objectives of the Department of Food, to the need for more accurate economic evaluation of loss prevention activities; the Department of

Food were keen on expanding immediately their Save Grain Campaign.

6. Organised by the Department of Food, Ministry of Agriculture, the Indian Agricultural Research Institute and the Institute of Development Studies in January 1981 and funded by UNU and GTZ.

7. Despite the fact that now there are many other figures available for which, unlike these, accurate information on sources and methods of data collection are available, they continue to be highlighted eg UNEP (1983).

8. See Greeley (1978) for a discussion of the different problems associated with the diffusion of improved traditional stores and of urban manufactured metal bins.

9. One of the most well known examples is the use of modern Swedish grain storage silos in Tanzania based upon the recommendations of a Swedish consultant engineering company; several commentators on this programme have drawn attention to the inappropriateness of these structures.

10. Canberra is the home of the Commonwealth Scientific and Industrial Research Organisation, Culham is the home of the Tropical Products Institute, (now the Tropical Development and Research Institute) Industrial Development Department and Kansas is the home of the Food and Feedgrain Institute of Kansas State University. All these organisations have been very heavily involved in farm level food loss prevention programmes and test and evaluate technology for such applications in tropical countries.

11. Gill also draws attention to the problems caused by the concentration of market arrivals which have tended to become more severe as the marketed surplus has increased. According to the results of his study in wheat and rice markets in the Punjab and Haryana, maximum daily arrivals increased ten-fold in Khanna market, Punjab State from 1965-67 to 1978-79 (Gill, 1982 p156).

12. The assumption seems to have been that if modern market facilities were the locus of serious food losses then traditional farm-level postharvest practices must be suffering catastrophic losses.

13. As Bergeret (1981) states the greater susceptibility of the HYVs is well known but this phenomenon is not true of all the new varieties; see IARI (1980).

14. For example it was observed in the course of field studies in Andhra Pradesh, India in 1974 to 1976 that farmers growing the rice variety RP 4-14 because of its high yield and fine grain quality were, almost universally, unwilling to store it because of the problem of insect

attack. They would plant all but a small part of their land to the HYV and would use that small part for a traditional consumer preferred variety which they knew they could store effectively.

15. These details were learnt in conversation with Paul Hindmarsh, a TPI expert then working with the Ministry of Food in Dhaka.

16. In India, partly with IDRC support, there was the all-India coordinated postharvest technology scheme and SEARCA operate a post harvest network for South East Asia with multiple donor agency funding.

17. Both in India and in Bangladesh there is evidence that decisions on rice milling technology have been taken for reasons other than economic viability and various types of commercial pressure have been important in determining these choices (Barbara Harriss, personal communication, 1978). There are also journals (so-called) such as Agricultural Mechanisation in Asia whose primary purpose, to judge by the articles, is to persuade developing Asian countries that modern rice processing technology, preferably purchased in Japan, is ideal for their rice economies; see for example the article by Satake (1978) on Bangladesh.

18. Shell, for example, in promotional literature for their product GRAMEX seem to be suggesting that use of their product will increase food availability by 10 per cent by preventing postharvest losses.

19. The Shilpa Bank in Bangladesh, responsible for financing private investment in modern rice mills, provides financial justification of that choice of technique almost entirely on the basis of the supposed extra 2 per cent head rice yield, compared to alternative milling techniques. (Personal communication, Mr. Rizvi, Shilpa Bank 1980).

20. These studies were: Harriss (1978), Bangladesh; Harriss (1974), India; Timmer (1974), Indonesia; Stewart (1978), Kenya; and, Spencer et al. (1976), Sierra Leone.

21. 'Bias' exists when, with constant factor prices, the direction of technical change does not save capital and labour in proportion to current factor shares.

22. Coursey (1981) gives an account of exactly the opposite process taking place where village level technology is being scaled up to 'industrial' levels because of its loss reducing characteristics.

23. In some circumstances the shortfall would have been met by borrowing grain which would have to be repaid with interest sometimes as much as 50 per cent even though the period of borrowing may have been as little as a month or two; in these cases the modification requires an increase

rather than a decrease in the net benefits of loss reducing technical change.

24. In the work in Bangladesh, farmers' estimates of moisture content based upon biting the grain were compared with the estimates given by moisture meter. Results suggested that farmers could identify very wet, slightly wet and dry grain but that the precise differences between 12 per cent moisture content and 14 per cent moisture content, the critical moisture range for safe storage, were not detectable by farmers in all cases. There was also frequent disagreement about the relative wetness of the grain.

25. It is to be noted that the predominant insect pests were Sitrotoga Cerealella which usually infest the surface of a grain store and therefore the fact of their presence did not necessarily indicate severe insect loss - that depended most upon the design of the grain store in question.

26. Of course it could also be the case that there was an insufficient appreciation of the economic benefits from use of the insecticidal dust because the economic returns appear to be quite substantial.

27. Farmers in Madhupur in Bangladesh reported that they continued to use hired female labour rather than use the rice mill because they had an obligation to employ them based upon a long standing relationship of employment.

28. Ownership of metal bins was a sign of affluence in these states and the farmers were proud to be seen coming back from the market with a metal bin on the back of their bullock cart.

29. The best known of these are UNIDO (1972) and Little and Mirrlees (1974).

30. The Consumption Rate of Interest is Little and Mirrlees (1974) term. A more widely used term is the Social Discount Rate (SDR); however, Little and Mirrlees use the term CRI to distinguish the rate at which future consumption is discounted from the rate at which future project income is discounted since project income will consist of both consumption and savings. This latter rate they call the Accounting Rate of Interest (ARI). UNIDO (1972) use aggregate consumption as their numeraire (see footnote 32) and in their method the CRI is in fact equivalent to the SDR.

31. A more detailed discussion would require qualification because workers do save and owners of capital do consume.

32. In fact the method of adjustment depends on the particular cost-benefit method being used. Little and Mirrlees (1974) use uncommitted social income as their numeraire, which, very roughly speaking (see Chapter Six below and Little and Mirrlees pp.145-151), is equal to savings. Other manuals (notably UNIDO 1972) use aggregate consumption as their numeraire; in this case the value of savings is adjusted upwards (by P_{inv}, the price of investment in terms of aggregate consumption) to express its equivalence in units of consumption. The difference is only one of method.

We can also note here that in the first adjustment described, distorted market prices, the new prices (accounting prices) are all expressed in terms of the numeraire. The adjustment for the price of savings which we have called the second adjustment is but a special and important case of the adjustment for distorted market prices.

33. Even if the output of the particular project in question was capital goods the effect we are concerned with is the effect on consumption goods - if the production of capital goods uses more resources than technical efficiency dictates it means less resources are available for production of consumption goods elsewhere.

3

Rice in Bangladesh: The Farm-Level Postharvest System

3.1 INTRODUCTION

The purpose of this chapter is to set the Bangladesh context for the discussion of food losses and technical change in later chapters. Following the introduction there are five sections. 3.2 is a 'what are we trying to measure' section that offers definitions of 'postharvest' and of 'food loss'. 3.3 briefly describes the seasons of rice production, the usual pattern of postharvest operations and the causes of rice postharvest losses in Bangladesh. 3.4 examines the planning context for farm-level postharvest technical change in the Bangladesh rural economy with special reference to the current five year plan. 3.5 reviews the literature available to the author on the Bangladesh postharvest system. It deals first with the quantitative estimates, all unmeasured, of food loss and then looks in turn at: the government postharvest Task Force report; socio-economic literature; studies on rural women and the postharvest system; and finally, with other, miscellaneous, postharvest studies. The final section, 3.6 uses field data to look at the end uses of rice in relation to the valuation of postharvest losses and at the use of labour and capital equipment in traditional farm-level postharvest operations.

3.2 WHAT ARE WE TRYING TO MEASURE?

'Postharvest' and 'Food Loss' are both compound expressions frequently the subject of uncertainty and disagreement. The term 'postharvest' is controversial because of difficulty defining which operation it begins with and where it ends. Whilst the widely quoted NAS (1978) view that 'the postharvest period of time begins at separation of the food item from the medium of immediate growth or production' is accurate and precise it does not reflect the fact that studies of postharvest systems usually start with the cutting operation. They do this for two related reasons. First, a major determinant of postharvest losses is the condition of the crop at harvest. For example, a crop which is allowed to mature longer in the field will be harvested at a lower moisture content; this reduces drying requirements but may increase the level of transportation losses due to shedding - of course, if shed grain is retrieved by others this is a loss to the farmer only and not to society. Secondly, losses in the harvesting operation itself are more closely related to the transportation and threshing loss interests of postharvest specialists than to the field yield interests of crop production specialists.

To circumvent this definitional problem some authors (eg de Padua 1976, Huq and Greeley 1980) have used the term 'post-production' instead. This term is more accurate but unfamiliar; therefore, reflecting general usage 'harvesting'losses have been subsumed within this 'postharvest' study. It should be noted that this broader definition of postharvest does not extend to pre-cutting losses due to rodents etc., nor does it include reduction or loss of crop output because flooding prevents cutting.

The NAS (1978) definition of the end of the postharvest system is 'when the food enters the process of preparation for final consumption.' This is less precise. Does dropping peas off a fork count as a loss in the preparation for final consumption? Generally, the postharvest system is taken to end, as NAS (1978) later states, before meal preparation activities begin. This does not of course imply that food losses during meal preparation do not occur or are not significant but only that they require different types of intervention (changes in eating and cooking habits) to reduce loss, and therefore provide a useful boundary. In fact, such losses can be substantial; (this thesis goes on to show that the estimated physical losses in cooking rice were on average greater than total physical losses from cutting up to cooking). Recognising that the definition, which is necessarily arbitrary to some extent, excludes these aspects of food loss, it is accepted here that the period from cutting to just prior to final meal preparation defines the postharvest system.

Food loss is difficult to define because of disagreement about what constitutes loss and because the different components of loss cannot be easily aggregated into a single index of food loss. The food value of quantitative and qualitative losses, the economic value of losses, the nutritional significance of food loss and culturally specific food habits are the main areas of debate, though there are others in special cases such as germinability. However, for a policy-oriented discussion of food loss it is possible to be reasonably precise concerning three requirements that an estimate of food loss must fulfill.

First, it must be accurately measurable. A degree of discretion, perhaps only a small amount, in determining the necessary accuracy of a measure is possible depending on the true level of losses and upon the productivity of possible alternative techniques. There are three situations where some rules of thumb can be used; if losses are very very low (below one per cent say in Bangladesh for a specific farm-level operation) then loss-reduction is unlikely to be a significant influence upon technical choice. With such losses even a coefficient of variation of 100 may not be reason for uncertainty over the cost effectiveness of alternative techniques. The middle range of losses (say 4 to 12 per cent) is the most critical because this is where their value is likely to be neither insignificant nor overwhelming. Differences of 2 or 3 per cent may be critical in evaluating alternative technical choices. The upper range (above 15 per cent) may not require precision if the minimum estimated value of true loss is above a level that makes alternative techniques preferable. If the productivity - labour and capital costs per unit of output - of new techniques is vastly superior then inaccuracy in the upper ranges is less likely to matter. However, the availability of more than one alternative, which themselves have different loss characteristics, limits the possible applications of this rule of thumb.

Secondly, the estimate must be soundly defined. Developing a set of operating rules for measuring losses, as stated in Chapter Two, has been identified as a critical need for improving choice of techniques analysis of farm-level postharvest operations. There are four areas where difficulty in providing an accurate estimate has been most evident. (a) Grain sampling methods and laboratory procedures have been inappropriate and equipment has been subject to shortage and misuse. (b) Frequently the rather obvious requirements, that an estimate must be specific by type and cause of loss and by where it occurs, have not been observed. (c) Even more frequently, the population to which the estimates refer has not been clearly specified. Very few loss assessment projects are sufficiently large that they can provide independent estimates for more than very crude categories of farm-level situations. The circumspection in proferring estimates that this fact should warrant is rarely encountered. Most calculations of national foodgrain losses at farm-level, if they have any empirical base at all, are based on estimates from small samples without indication of statistical confidence limits. (d) Bias in estimation procedures often results in mean estimates higher than the true mean; this is not due solely to the possible subjective bias of the research team but to the non-representativeness imposed on sampling procedures by conditions of field work (see Section 4.5). Since the modal value of losses will always be lower than the mean value (the distribution is skewed to the right), the mean value will anyway always overstate the most likely level of loss for any one farm; reported means may therefore grossly exaggerate the actual farm-level situation.

Thirdly, the estimate of loss must be translatable into an economic value. This condition is basic to analysis of alternative techniques but it does create some constraints on what can be accommodated within an estimate. (a) Most fundamentally the economic value basis divorces the estimate of loss from the true nutritional loss to the extent that qualitative deterioration in nutritional worth is not directly reflected in economic worth. It is possible that nutritional and economic values will actually move in opposite directions, for example if there is a price premium on highly polished rice. (b) Culturally specific food habits define what constitutes food; food loss is therefore similarly defined. Broken rice may not be a loss for a poor subsistence farmer even though it would be rejected or priced down in a supermarket. Quantitative losses are usually relatively straightforward to value but qualitative loss is far more problematic (see Section 3.6). (c) Whilst the economic value condition provides a basis for direct comparison of quantitative loss and some types of qualitative loss it explicitly excludes qualitative loss for which an economic value cannot be calculated.

These constraints introduce a potential difference between social and individual concepts of loss. For example, quality characteristics, which are not observable when grain is sold but which are measurable, can be given an economic value to society if they also constitute nutritional loss even when they are not a cost to the farmer. One such case would be loss of thiamine during wheat storage. This raises a difficult question. Should the social value of food loss exactly equal the nutritional loss? More precisely, should the rupee value we attach to a particular occurrence of qualitative food loss be equivalent to the rupee replacement cost - assuming such a cost could be calculated - of the lost nutrients? If the sole purpose of eating were to stay alive the answer would probably be yes, at least if the 'loser' was undernourished. But, meals are social occasions and gastronomic events that have value other than mere survival. Furthermore, even if food was merely nutrition, as some might reasonably argue is a correct view where hungry people are involved, the nutritional implications of specific food loss situations would be very difficult to define. There have been attempts (Gupta 1981) to prepare a single index of food loss that weighted different nutritional components but they are unsatisfactorily imprecise. They also imperil the measurement and sound definition criteria.

Aspects of qualitative deterioration that are of social significance but not reflected in economic value criteria must be dealt with but their incorporation into a single food loss index is therefore not possible. An important example would be where mycotoxins go undetected and therefore do not affect economic value but are sufficiently serious to cause sickness or death. In such cases, separate calculations on 'loss' prevention needs are required. It is of course often feasible, and always desirable, that the research team providing estimates of the economic value of losses could provide evidence in such cases.

A final point on the economic value of losses is worth making here. It is conventional to value losses from different stages of the postharvest system at the same price; for equal quality grain the value conventionally attached to a loss of fifty kilograms during harvesting is the same as that attached to a loss of fifty kilograms during threshing. Strictly speaking, this practice is wrong. Unthreshed wet paddy sitting in field bundles is not as valuable as threshed, dried and cleaned paddy sitting in bags in the farmyard. The value attached to harvesting losses ought to be net of the additional value added in effecting the sale. The validity of this point becomes clear when considering how losses of paddy and rice would be valued. Suppose a mill converts 100 kilograms of paddy into 72 kilograms of rice. The value of those 72 kilograms of rice would be greater than the 100 kilograms of paddy in proportion to the milling costs, and this is always taken into account. There is no good reason why a similar adjustment is not made for valuing losses at earlier stages; we suspect the reason is either that the difference is considered to be too small to warrant an adjustment or simply that it has never been thought about. We do not pursue this point any further except to note two implications. First, that the economic value of losses at the earlier stages of the postharvest system are overstated by simple use of market prices. Secondly, that the economic incentive to the farmer to achieve a given saving will be greater at later stages in the postharvest system than at earlier ones.

In summary, measurability, sound definition and economic worth criteria are the essential components of a loss estimate. In Bangladesh, the major part of production is for self-consumption. For that grain, qualitative deterioration that does not cause a change in final use may not constitute loss (see discussion in 3.6). (When a final use change does occur the difference in value between the uses is the correct measure of loss.) For the marketed surplus, quality considerations will obviously be relevant though in poor societies where quality price premiums are sometimes relatively small they will be less significant than in richer societies which can support a wider range of consumer preferences. In practice for poor subsistence producers, quantitative loss measured by weight loss will be the major component of food loss with qualitative loss becoming more important as economic penalties for poor quality grow. This approach has support from nutritional studies which have established unequivocally that for most poor consumers inadequate calorie intake is the cause of protein malnutrition and that the quantity rather than quality of food is the critical consideration in combating malnutrition (Sukhatme 1972).

These definitions of postharvest and of food loss are informed by very practical considerations and will not satisfy everybody. Bourne (1977), for example, defines postharvest as the period between 'the completion of harvest and the moment of human consumption'; and he defines food loss as 'any change in the availability, edibility, wholesomeness or quality of the food that prevents it from being consumed by people'. Thus anything originally designated as food and subsequently used as feed because of deterioration is a 100 per cent loss, despite the positive economic value of the feed. Another view, that Professor Hans Singer has suggested in discussions, is that postharvest losses should include reductions in a person's utilisation of nutrient intake because of disease and malnutrition. Development of these different perspectives on the nature of postharvest food loss usefully draws attention to the multiple dimensions of food and nutrition problems. However, these dimensions are not always of immediate operational significance, in part because they are insufficiently understood and in part because they do not relate to specific intervention strategies implementable by the planners and scientists concerned with postharvest technology. The development of postharvest policy, initially at least, must depend upon a more practical concern with what can be readily measured and can be compared with realistic alternative practices, loss-reducing or otherwise. This is the justification for the use here of physical food loss as the main subject of research on losses, supported by analysis of qualitative deterioration for its economic implications.

3.3 CROPPING PATTERNS AND POSTHARVEST OPERATIONS

About eighty per cent of the gross cropped area in Bangladesh is under rice cultivation (estimated from GOPRB 1983a, pp.167 and 180). According to the available statistics both the gross cropped area (1982, 3.2 million acres) and the acreage under rice (1982, 2.6 million acres) have been growing marginally every year over the last decade and the rice acreage has been a fairly constant proportion of gross cultivated area. The seasonal distribution of rice production is summarised in Table 3.1, together with the proportion of each season's area under HYVs. The physiographical and climatological conditions influencing the pattern of rice production have been described in detail *inter alia* in Ahmed (1976), in the cost of cultivation surveys of the Ministry of Agriculture and Forests (GOPRB 1979) and in the references provided there. The Bangladesh Rice Research Institute with the International Rice Research Institute have published a detailed agro-climatic survey of Bangladesh (Manalo n.d.) which presents a clear picture of regional and seasonal variations in rainfall, temperature, wind and hours of sunshine, relative humidity and evaporation; this provides a useful basis for understanding the influence of climatic conditions on postharvest

Table 3.1
Seasonal Distribution of Rice Production (1978-79)

	Aman	Aus	Boro	Total a/
Total Area ('000 acres)	14,347	7,995	2,649	24,991
% of total rice area	57.4	32.0	10.6	100
Total yield ('000 tons)	7,429	3,287	1,929	12,646
% of Annual Yield	58.8	26.0	15.2	100
% of Area				
local varieties	88.1	86.8	37.7	82.3
pajamb/	5.7	0.4	6.3	4.1
hyvs	6.2	12.8	56.0	13.6
% of Yield				
local varieties	79.9	70.8	21.7	68.6
pajam	8.8	0.5	6.8	6.3
hyvs	11.3	28.7	71.5	25.1
Yield per acre (tons)				
local varieties	0.47	0.34	0.42	0.42
pajam	0.80	0.63	0.79	0.79
hyvs	0.95	0.92	0.93	0.93
area irrigated (%)	2.2	2.9	91.0	11.84
area broadcast (%)	29.0	n.a.	0.0	n.a.

Notes: a/ The total row may not correspond to the aggregate of the season totals due to rounding.
b/ Pajam is an improved variety that has a seperate classification in official statistics.

Source: Compiled from GOPRB, 1980b, pp 20-55 except area irrigated which is derived from GOPRB, 1983a p.197.

operations and on losses. There are three points suggested by this literature that are of special importance for a postharvest study. First, the likelihood of prolonged rain at the time of harvest is the most important determinant of the risk of postharvest problems; two harvesting seasons, the boro and the aus are susceptible to prolonged rain. The boro crop was traditionally a lowland local variety grown using traditional irrigation equipment or where water remains after the annual floods have receded. The introduction of more and better irrigation facilities is extending the area under boro and encouraging the use of HYVs which by 1982 accounted for over 60 per cent of the total sown area in boro. It is a transplanted crop, and particularly with the later maturing HYVs, is harvested at the beginning of the monsoon. Aus is grown in highlands under rainfed conditions, and sometimes now with irrigation from shallow and deep tubewells; it is also grown on medium highland where flooding provides the water and in these cases it is sown mixed with an aman crop. In the first of these patterns the crop is sometimes transplanted, though usually not, and only about one fifth of the area is sown with HYVs. In the second pattern, the crop is always broadcast and only local varieties are grown. Harvesting takes place in June through August with the early aus difficult to distinguish from late boro. This is the peak of the monsoon and the main focus of postharvest research (see 3.5 below) has been on the development of suitable drying facilities for these crops.

Secondly, the main opportunity for increase in rice production in Bangladesh lies in the use of HYVs during boro. In much of the aman area, the main rice season, including areas where transplanted aman is grown, the risks of cultivation have restricted the use of HYVs to less than 10 per cent of the area. In some regions, triple cropping (broadcast aman followed by two rabi crops) is being replaced by a single boro crop; more commonly the boro crop is followed by a transplanted aman crop and occasionally by both an aus and aman crop. The effect of the emphasis on boro cultivation is to increase the proportion of annual output which is exposed to wet season postharvest problems.

Thirdly the growth of the boro acreage is making the timing of production operations more critical. In particular, the harvesting of irrigated boro, and sometimes aus, is most commonly followed by the transplanted aman crop and is heightening the seasonality of farm work. The demand for labour and bullocks to be used in preparing the land and sowing the transplanted aman crop - accounting for 40 per cent of total cropped rice area - coincides with increasing demand for those same resources to be used in postharvest operations on the boro crop. This is an important consideration in estimating the opportunity cost of 'best' postharvest practices from a food loss perspective and is discussed in some detail in Chapter Five in the comparison of pedal threshers with traditional threshing methods.

The postharvest practices traditionally followed in Bangladesh vary very little by season and with few exceptions by region. The usual sequence is as follows.

1. Cutting by sickle and field stacking; where flood waters have not receded the crop may be placed in a boat or where the field is very dry, spread out on the stubble for drying. Most commonly it is stooked, bound by rice straw and kept in the field or on the field edge for transporting to the farmyard. Physical losses occur because of shattering and, in the deepwater fields, because of uncut plants; the moisture content sometimes increases rapidly when the cut stacks are laid on wet ground.

2. Transportation of the stooks is usually done on the head or by shoulder-pole; boats and bullock carts are less commonly used. Occasionally, poor farmers have been observed laying the harvested crop in a long line of stooks and floating the crop home through ditches and canals. The only cause of loss in this operation is further shattering.

3. Farmyard stacking involves a period varying from a few minutes to several days. The stack is usually in a bee-hive shape with the heads inwards. This results in very high temperatures (sometimes, but rarely causing combustion) and the formation of moulds and some germination around the edges of the stack. Extended stacking occurs because of resource constraints (threshing labour/animals being used in ploughing/sowing for the next crop) and because of rain preventing threshing.

4. Threshing and Winnowing;threshing is completed through hand beating, usually on a drum or a piece of wood, or bullock treading; sometimes both methods are used with a second period of stacking between the two. Some physical damage can occur because of the stress on the rice grain as it is separated from the straw but the main source of loss is incomplete threshing. Some commentators argue that grain lost in the soft floor of the threshing yard is an important source of loss but,

as discussed below, this is not supported by field evidence. The threshing operation is one of the two operations where technical change is occurring, with the introduction of pedal threshers. Winnowing is done using a tray (a kula): the grain and chaff are carefully thrown up into the wind which separates the grain from the chaff. This is not a significant cause of loss.

5. Soaking and Parboiling are practised in all but the Eastern and Southeastern districts of the country. Gariboldi (1974, pp.4-5) lists 11 advantages of parboiling which is literally partial cooking of the rice whilst the husk is still on. There are major differences, region to region, in the period of soaking and in the length and number of times of parboiling. The grain is usually soaked in large earthenware, aluminium or metal pots (Raings), or in sacks in a pond allowing water to penetrate the grain. It is then steamed over a fire for a period between an hour and three hours. This process gelatinises the starch making it hard, more nutritious (see Gariboldi 1974, p.5) and imparting a distinctive taste, texture and colour to the rice. Parboiling reduces milling losses due to fracturing of the rice by sealing such fractures but compared to raw paddy, parboiled paddy is more susceptible to some types of insect attack because the two sides of the husk (the lemma and the palea) are split facilitating oviposition, particularly for the lesser grain borer, (Rhizopertha dominica), (Breese, 1963).

6. Drying takes place both on the threshed raw paddy and parboiled paddy; occasionally paddy from store is further dried prior to milling. Drying is by the sun on prepared mud surfaces and occasionally - though it is not nearly so significant as a tarmac-based expedition into the countryside might suggest - on hard road surfaces. The main cause of loss according to most analysts of Bangladesh postharvest operations is the absence of mechanical drying facilities and the delays so caused because of rain preventing drying (see, however, Section 4.8 below). Two further possible sources of loss are grain lost in the drying floor - though as with the threshing floor losses field evidence does not support this - and reductions in milling yield because too rapid drying causes grains to fracture. Qualitative losses (malodour and discolouration) are frequently

mentioned aspects of drying problems and the economic evaluation of qualitative deterioration can be difficult especially when the crop is not marketed.

7. Storage at farm-level in Bangladesh has been described in detail in Boxall et al 1976; the main structures used are bamboo or reed baskets coated with a mud and dung mix, earthenware pitchers and sacks; metal drums, straw structures and separate rooms (mud-built) in the farmhouse are also used but in fewer cases. The main causes of loss are insect and rodent attack but seasonal increases in moisture content can result in mould formation and reduction in the quality of milled rice.

8. Husking and Polishing (milling) are done in two main ways. Traditionally, the *dheki*, a foot-operated pestle and mortar has been used; the friction of the pestle against the paddy in the mortar causes a shearing action which separates the husk from the rice and in polishing removes the bran layer. Whilst the rice yield is high by this method much rice is broken when it is trapped at the bottom of the mortar and gets smashed.

Huller rice mills of the Engleberg (also known as Kiskisan) type are now replacing the *dheki* to varying degrees in different regions, strongly influenced by the availability of electricity. Araullo et al. (1976, p.292) describe them as follows. "The Engleberg huller removes husk and bran in one operation. Paddy is fed into one end of a hollow cylinder containing a rapidly rotating fluted roller. Axial movement of materials is accomplished by (i.e. paddy moves from the inlet because of) an integral, truncated screw conveyor located at the point where the paddy enters the mill. Husking is the result of the shearing action on the grain produced by the movement of the roller flutes past a stationary blade. The combination of rapid rotation, high pressure and inter-grain contact results in removal of both husk and bran as material moves axially through the machine. Husks, bran and broken rice pass through the perforated screen in the lower half of the cylinder while milled grain is discharged at the terminal end of the mill" (Araullo et al., 1976, p.292).

The huller does not increase the quantity or quality of rice but considerably reduces the time taken in husking and polishing. It is the second main technical change occurring in postharvest operations in Bangladesh and is the subject of Chapter Six. The introduction of large modern rice mills and of medium size rubber roll shellers is also occurring but these techniques are used for the marketed surplus by traders and the public sector and have little effect on farm-level operations.

3.4 POSTHARVEST TECHNICAL CHANGE AND PLANNED RURAL DEVELOPMENT: AN OVERVIEW

Employment-intensive agriculture-led rural growth represents the core of Bangladesh's development strategy. The Second Five Year Plan, 1980-85 (GOPRB 1980a), provides a detailed specification of the approach and places particular emphasis upon the food self-sufficiency objective and the generation of rural employment. In Chapter Two, the relationship of these objectives to postharvest technical change has been discussed in general terms and the following paragraphs relate the issues raised there to the specific conditions of Bangladesh.

The conditions determining Bangladesh's development strategy are well publicised. Table 3.2 summarises some of the more recent statistical evidence. Of particular note are the four estimates (1975 or 1976) of the rural poor, defined as those not meeting nutritional needs, as a percentage of the total rural population which are 59, 70.3, 71 and 84.2. These differences require some comment.

All four estimates use an age, sex and activity group (ASAG) adjusted computation of per capita caloric need. The Ahmad (1977) estimate is based on a survey of actual dietary intake involving one day of recall and one day of actual weighment for each of 699 households; the other three estimates are all based on the 1973-74 Household Expenditure Survey. It is unusual then that they obtain such a wide range of estimates. Even more unusual is the fact that the four ASAG-based estimates of calorie requirements were all different and with the exception of the IBRD - Osmani comparison, the numbers of poor people were inversely related to the numbers of calories specified as the poverty line; Ahmad (2248 calories and 59 per cent); Khan (2150 and 70.3); Osmani (2085 and 71); and IBRD (2122 and 84.2). The fact that Ahmad's estimate of numbers of poor people is much lower despite use of a higher calorie standard could be because of small sample size or because of seasonal factors - the months of September, October and November were not included and they are known to be times of food shortage, (Chowdhury et al 1981, Becker and Sardar, 1981, Chaudhury, 1981). Alternatively, the three estimates based on

Table 3.2
Bangladesh Statistics Related to Rural Poverty

GNP per head a/	140	US dollars
GNP growth rate a/	0.3	per cent
GDP from agriculture a/	47	per cent
Total Population b/	92.6	million
Rural Population b/	80.4	million
Population Density b/	1665	per square mile
Life Expectancy a/	48	years
Infant Mortality Rate a/	133	deaths before age one per 1000 live births
Total Fertility Rate a/	6.3	expected live births per woman
Population Growth Rate a/	2.6	per cent
Literacy (5 years and above) b/	26	per cent

	Percentage of Rural		
Crop Land Owned (1978)c/	Households	People	Land
Landless	28.78	23.43	0.0
0.01-1.00 acres	33.20	30.33	8.41
1.01-2.00 acres	14.55	14.95	13.27
2.01-3.00 acres	7.81	8.88	11.89
3.01 acres and above	15.66	22.41	66.43
Crop Land Leased In c/ (1978)	35.47	-	23.65
Landless Leasing In c/ (1978)	7.41	-	4.69

Rural Population below Poverty Line (1975-76)		
Ahmad (1977)	59	per cent
Khan (1977)	70.3	per cent
Osmani (1980)	71	per cent
IBRD (1980)	84.2	per cent

Note: All data is for 1982 unless otherwise specified.

Sources: a/ = IBRD 1984a pp218, 222, 254, 256 and 262.
b/ = GOPRB 1983a p88
c/ = GOPRB 1978a Tables I and IV

expenditure could be overestimates because of failure to adequately account for consumption from own-production. The differences between these three estimates, which all share a common source for expenditure data, is due to methodological differences largely relating to assumptions about sources of calories, prices and disaggregation from the expenditure groups used in the original data.

It is not possible to judge the relative merits of these estimates. However, even the lowest estimate suggests three fifths of households are malnourished, a figure so high that the validity of any of the poverty lines is questionable; any plausible distribution of households below the poverty line would involve a high percentage of rural households in sustained negative energy balance. In fact, two of the authors cited above recognise this and provide population estimates of the hard-core poor, 60 per cent (IBRD 1980, p.199) and of extreme poverty, 41 per cent (Khan 1977, p.147) based respectively on 85 and 80 per cent of their minimum caloric requirements.

This problem of interpretation of the poverty line has been discussed for some time in relation to Bangladesh (see Bose 1968, p.473-4 and Rahman 1970 cited in Islam 1977, p.67): Rahman comments 'Postulate any reasonable estimate of minimum subsistence needs and we can point out a significant number of persons, in any developing country consuming less than this minimum and yet very much subsisting'. The estimates in the table for 1975-76 compare with 1963-64 estimates of 36.7 per cent (IBRD 1980, p.199) and 40.2 per cent (Khan 1977, p.147) in 1963-64, since when they have been more or less continuously increasing. The poverty line estimates in the table must therefore be interpreted not as a precise measure of malnutrition, for they either underestimate consumption or overestimate need, or both, but as a murkily defined indicator of growing poverty.

The growth of poverty is associated with falling real wages (Khan 1977, p.151, IBRD 1980, p.203) and the growing incidence of rural landlessness (Khan 1977, p.155, Alamgir 1978, p.101). High population growth rates (GOPRB 1981, p.3) and declining food availability per head (GOPRB 1983a, p.524) are also associated with this trend of increasing poverty. With 87 per cent of the population in rural areas (GOPRB 1983a, p.88) and 50 per cent of the rural population functionally landless1/ (Jannuzi and Peach 1980, p.110), the Second Plan's emphasis, predictably, is on rural employment generation and provision of cheap wage goods for poor rural consumers. In practice this involves growth in agriculture both as the source of wage labour opportunities and of the wage good (food) that rural consumers spend 75 per cent of their income on (IBRD 1980, p.191).

The Plan proposes an increase of employment from 25.3 to 30.5 million man years. Of the 5.2 million increase 3.2 will come from agriculture (GOPRB 1980a p.V1-7), an increase of 16 per cent. It is difficult to interpret this estimate in terms of its effect upon the level of unemployment because participation rates and the level of underemployment are uncertain. The Plan acknowledges these difficulties and cautiously estimates that the labour force will grow by 3.8 million man years during the plan period, due to population growth; without any change in participation rates, the net increment to employment will be 1.4 million man years. The plan document is extremely cautious both concerning the likelihood of achieving the employment targets - calling them 'feasible limits' rather than projections - and the effects that doing so may have on the level of unemployment (GOPRB 1980a, p.V1-11). The achievement depends upon an overall agricultural growth rate of 6.3 per cent and a principal foodgrains (rice and wheat) annual growth rate of 7.2 per cent when in the past it has never been sustained at more than 2.7 per cent (Hossain 1980, p.39) and was negative in the year prior to the Plan (GOPRB 1983a, p.270).

Before turning to the possible implications of this approach for farm-level postharvest technology, it is necessary to note four, more political, considerations which determined the direction of planning and which all relate to food availability and employment implications. First, as with other key targets, notably employment, the adoption of a food self-sufficiency target at a higher level of consumption was a political imperative that dictated the agricultural production strategy proposed; the target has been raised from 15.4 to 17.2 ounces of foodgrain per person per day - actual consumption in 1979-80 was about 14.8 (337 lbs per year GOPRB 1980a V-6). Throughout the document it is apparent from the many cautionary notes that the sectoral planners are not convinced of the realism of the targets set. The plan document actually contains three different targets for crop production during the plan period; one based on the food self-sufficiency objective, one on the employment objective and one on a 1985 target production rate if a doubling of food output is to be achieved between 1980 and 1987. This last target involves more than a one third increase in output (an extra 2 million tons) compared to the food self-sufficiency objective (GOPRB 1980a, p.XII-40 (a)).

The employment determined crop production objective (GOPRB 1980a, p.VI-12/13) requires an additional growth of 2.8 per cent annually on top of the food self-sufficiency objective to meet the employment target. This represents more than a 45 per cent increase in the growth rate of crop output. It is presumed this will occur through more intensive use of traditional irrigation, consolidation of holdings, changes in tenancy arrangements and similar hopes.

Whilst the political emphasis upon rural employment through increased crop production is supported by the highest ever planned allocation of one third of public sector outlay to agriculture2/ and total rural development expenditure takes 53 per cent of total planned outlay, the changes in cropping patterns, acreage increases and input supply requirements are far in excess of past performance. Mahmood's (1980, p.9) comment that the Plan looks like a telephone directory and reads like a Holy Book is not without relevance.

Secondly, the institutional basis for the proposed strategy is uncertain. The role of the village councils and of the IRDP remain particularly confused. As the Plan euphemistically states (GOPRB 1980a, p.VII-2) 'the last few years, particularly the last 2-3 years, have been a period of experimentation with several methods of rural development'. The institutional strategy that has emerged, termed Comprehensive Rural Development, though heavily qualified (GOPRB 1980a, p.VII-5/6), involves the creation of a new tier of local government at the village level. The arguments in favour of this plan have not been entirely convincing, particularly regarding the possible impact upon the poorest even though they are supposed to be principal beneficiaries (Chowdhury 1980); that is another subject, however, and the more simple argument advanced here is that the very fact of radical changes in rural institutional structure will involve a disruptive transitional phase that will necessarily impede the implementation of the Plan objectives in the immediate future.

Thirdly, whilst at some points the Plan (GOPRB 1980a, pp.XI-6/7 and XII-104/105) explicitly endorses land reform including legal measures to acquire land over the new ceiling (unspecified), elsewhere this concept receives indifferent treatment. On Page II-6 in fact precisely the opposite is stated, viz. 'It is not however necessary to interfere with the ownership of land' and pages VI-11, VII-8/9 and XII-104, seem to suggest that other things may happen. In fact, at least so far, and the Plan is now more than three years old, no reforms have been attempted.

Land reform has a long and familiar history of advocacy without implementation; the issue the land reform question draws attention to is not just the question of land distribution but the underlying philosophy of the Plan to promote growth as rapidly as possible through private sector incentives without any attempt at direct income redistribution.

Improved food production and greater employment will definitely benefit many poor households, but given the land distribution (Table 3.2) their share in total output will necessarily decrease; the existing inequalities become more sharp as development proceeds because the policies upon which they are based represent a reinforcement of existing policies which have already proved to be inequitous (Asplund and de Vylder 1979).

Finally, and briefly, the Plan document recognises that food self- sufficiency has a sting in the tail; rural employment growth depends upon expansion of rural public works, particularly food for work which in turn depends upon food aid, estimated in 1979-80 to be worth $M 394 (GOPRB 1980a, p.I-18). Achievement of food self-sufficiency is explicitly concerned with removing this dependence but as the Plan states 'the prospect of such activities (rural works programmes) in the wake of attaining self-sufficiency in food and consequent stoppage of food aid does not seem very bright' (GOPRB 1980a, p.VI-17). The Plan requirement is for rural public works at a level 16 per cent greater than the pre-Plan period and by the end of the Plan to be financed entirely from domestic resource mobilisation. It is not difficult to envisage that for some rural consumers - the poor landless in backward areas - the effect of national self-sufficiency will be to heighten food shortage.

Together, these four considerations on the implementation of food availability and employment objectives suggest that the Plan may easily result in a worsening of the condition of the rural poor. Unrealistic targets, weak institutions, worsening inequality in land and strong pressure to reduce the food for work programme may all bring this about. This is the framework in which postharvest technical change has to be considered.

The Plan itself reflects little consideration of the role of postharvest technical change. The issue of farm-level food loss is not discussed anywhere. The introduction of improved postharvest techniques at farm level is referred to three times: '... pedal threshers ... will be manufactured in large numbers and sold to the farmers at reasonable prices ... (and) small paddy driers will be experimented (with) to test their acceptability to the farmers' p.XII-18; 'Facilities for mechanical threshing, drying and storing will be developed' (for aus paddy) p.XII-24; and 'processing of paddy by dheki which generates considerable female employment in the rural areas particularly amongst the distressed class would be encouraged by active research in improving the efficiency of (the) dheki to make it relatively more competitive than at present, so that there may not be any need for establishing new automatic rice mills3/ except for exports' p.XVIII-34; this last comment is in the seven pages devoted to women, and repeated on p.VI-17 in the section on employment where women are identified as a special target group along with the landless and the 'youth and educated'. In addition there are various references to the handling and processing of the marketed surplus (including the installation of crop driers at procurement centres to prevent postharvest losses), to the role of rural industries in supporting agricultural development through appropriate techniques and one reference to the development of postharvest technology by the agricultural research institutes. No specific proposals are detailed to support these statements; in some cases this is because the level of expenditure is too low to merit separate consideration4/ and in others because no actual proposal exists.5/

Although food-loss reduction, the principal concern of postharvest specialists, receives no attention the treatment of postharvest technical change does indicate four perspectives which are central to the concern of this book. First, and perhaps rather trivially, the technical changes discussed all relate to rice; rice accounts for about 80 per cent of the gross cropped area (GOPRB 1983, p.167 and 180) and any widespread change in rice postharvest techniques will therefore have a large effect. Even if rice is not severely affected by food loss and rice postharvest technical changes involve relatively small changes in factor use, the aggregate effects will almost always outweigh the possible effects of changes with any of the other crops which together account for about 20 per cent of gross cropped area.6/

Secondly, and in contrast to the employment optimism evinced elsewhere in the Plan, the postharvest employment implications of the proposed growth in food output are not considered. Certain postharvest operations - up to threshing generally - are included in production-related employment statistics but the largely family female employment in postharvest work is omitted; the dimensions of pre- and postharvest labour use are discussed in Section 3.6 below and here it is sufficient simply to point out that this is a significant omission but one that conforms to the more general under-enumeration of female participation in productive activities (see Greeley 1983a, p.35).

Thirdly, the Plan reference to pedal threshers is made in the context of technical change associated with increased productivity. Though not discussed solely with regard to these machines the introduction of labour-saving agricultural machinery is considered to be consistent with employment objectives when it releases a constraint on yield increases or when it is applied at the time of peak labour demand. As Chapter Five below, argues, both these conditions are particularly important in evaluating this technique.

Fourthly, the effects of technical change in rice milling upon the employment opportunities of the poorest women are raised. In fact, for the majority of women the introduction of rice mills that replace their family labour in rice husking and polishing represents a welfare increase but the Plan makes no mention of this. Nor should it be unduly criticised for the omission because by considering the effects on wage labour women it focuses attention on what is becoming recognised as a major social cost of modernising postharvest technical change. This theme is the one underlying the arguments developed in Chapter Seven of this thesis which attempt to substantiate the stronger claim that the provision of alternative income-generating activities for wage labour women displaced by rice mills represents one of the clearest and most direct opportunity to address the income needs of the poorest households in Bangladesh. Whilst the argument, does not support the case for improvements to the <u>dheki</u> as advocated in the Plan, the concern with female wage labour displacement is fully justified.

3.5 A REVIEW OF THE FARM-LEVEL POSTHARVEST LITERATURE

This review first examines how the Bangladesh farm-level postharvest literature provides support for the arguments developed earlier that food losses are generally presumed to be high and that loss-preventing technical change is presumed to be the priority for postharvest research. This is followed by a review of eight research programmes in Bangladesh which were concerned with loss reduction. It then examines important exceptions to these perceptions; <u>some</u> postharvest literature has addressed the economics of farm-level technical change and <u>some</u> has stressed the uncertainty about the levels of farm level food losses and the need for better estimates. Other analyses deal with technical change in relation to employment and income distribution, effects upon women, institutional obstacles to diffusion and energy implications. The debate in Bangladesh has not been completely one-sided and although this may not be the only reason why Bangladesh has not made the error of having a major farm-level 'Save Grain Campaign' as in neighbouring countries, it does increase the chances of field research results influencing the planning of postharvest technical change.

The High Food-Loss Lobby

In Table 3.3, 12 estimates of postharvest loss have been listed which range up to 100 per cent. The last of these (ref.12) is based on a questionnaire canvassed to 80 farmers in Mymensingh district during the 1980 aus season. The results form part of a more substantive study (Ahmed 1981, 1982, Ahmed et al 1980) that was undertaken at the same time as the research reported in the following chapters and there was some valuable informal collaboration. The authors are careful to point out that their results cannot be generalised because of the exceptionally clement weather prevailing during their farm-level work. Their results are rigorously presented with full details of methods, statistical significance etc. There is no doubt that this is the most thoroughly prepared and executed of the twelve studies discussed here and it is noteworthy that it is the only one to conclude that technical change to prevent food loss is not supported by the evidence.

Turning to the other eleven estimates, five of them (refs.6-10) were also based upon farm-level questionnaires. In one case (ref.7) information on sample size is not available. In the other four cases they were 30 (ref.6), 1 (ref.8), 180 (ref.9) and 50 (ref.10). Reference 8 was a study on a government seed farm; the two small samples were district specific and the large sample was a nationwide survey with between 5 and 15 replies from each district. Of the rest, only one estimate appears to involve some loss measurement; Hurley et al(ref.4) state (Appendix p.2) in relation to their estimates that 'very few of the figures have actually been measured, and where they have been measured they are confined to a few years, a few places and a few varieties'. It is not stated where these measurements come from or what they were. The other five estimates are all guesses.

Table 3.3
Estimates of Bangladesh Farm-Level Postharvest Food Losses

	Source	Estimate	Type of loss	Operations/Causes Covered by Estimate
1.	Satake (1978, p.42)	30-35a/	Not given	Postharvest operations
2.	Bala (n.d.but c. 1978, p.1, p.3	30-50	Not given	Lack of drying facilities
		25-30	Not given	Lack of drying facilities
3.	Samajpati et al (1978, p.1)	10-20	Not given	Processing and storage
4.	Hurley et al (1980, appendix)	5-44	Weight	Before Harvest to Washing and Cooking
5.	Sarker (1976, p.2)	30	Spoilage	Storage
	Sarker (1980, p.1)	10-25	Not given	Lack of drying facilities in *aus*
6.	Karim and Rashid (1979)	51-100	Premature Germination Decomposition of paddy. Formation of moulds	Due to the drying process about 70% of farmers lost the quality of their *boro* grain to the extent of 51 to 100 per cent

7. Task Force on Appropriate Technology cited in Farouk (1975, p.157)b/	22	Quantitative	Shattering during and after Harvest, Threshing, Drying, Storage, Handling and Transportation
8. Molla (1972)	22.7	Economic Loss	Lack of Drying Facilities for aus
9. Agricultural Maketing Directorate (1970, p.7)	7.95	Quantitative	Harvesting, Transportation, Threshing, Drying and Storage
10. BARD (n.d.)	10.17	Not known	Harvesting, Carrying, Threshing, Parboiling, Drying, Winnowing, Storage and lack of Drying
11. Faroukc/ (1975, p.157)	10	Quantitative	Postharvest losses
12. Ahmed et al (1980) p.16	0.82	Quantitative	Cutting, Carrying, Stacking, Threshing, Parboiling and Drying in aus

Notes: a/ Satake does not say whether it is only farm-level operations he includes although these are the only operations he discusses.

b/ Farouk cites the Dacca Division Task Force on Appropriate Technology but does not give a full reference. This paper was therefore not available to the author. Whether Handling and Transportation loss (6%) is only farm-level is not clear; a farm-level questionnaire was used.

c/ Farouk reviews earlier estimates (7, 8 and 9 in the Table) and concludes (p.157) that '10% may safely be taken for planning purposes'.

The background to the estimates by the Task Force on Appropriate Technology (ref. 7), by Molla (ref. 8) and by BARD (ref. 10) is not known but the remainder of these eleven studies can be divided into two groups. The first group (refs.9 and 11) are estimates in which the authors have no obvious or immediate interest in arguing that losses are either very high or very low; one is an inter-departmental government study group and the other is a literature review by an academic.

Other than the study by Ahmed et al these are the two lowest figures presented. The second group (ref.1-6) are all from people who have an interest in persuading others that the food loss problem is grave. The first author, Satake, is Vice-President of a corporation that is almost certainly the largest supplier of rice processing equipment to Asia. References 2 to 5 are all engineers working on drying (refs.2, 4 and 5) or storage (ref.3) projects. In two cases the estimates were used as arguments in research proposals. In another case (ref. 3) the research proposal said losses were 10 per cent and the research report, the application having been successful, said they were 10-20 per cent even though the research included no loss estimation work and no farm-level work. Karim and Rashid (ref.6) did their research in a village where they were organising an extension project to promote improved village technology where they had installed a rice drier. This second group thus constitutes an advocacy group - referred to hereinafter as the lobbyists - and this is clearly reflected in the fact that they provide four of the five highest estimates of loss.

All the lobbyists recommended adoption of new technology to prevent food loss. Half of them also recommend research to establish more precisely what losses actually are but this apparent inconsistency does not affect their identification of which operations need loss-reducing technology. Improved threshers, storage structures and drying facilities are specifically identified. The language is often unguarded; 'Modernisation of postharvest operations is imperative' (Satake 1978, p.42); 'Proper grain drying research would yield results which could ... minimise the huge losses of food and seed grain that are spoiled every year in Bangladesh' (Sarker 1976, p.2); 'An estimated loss of 25 to 30% of Aus crop occurs due to lack of drying facilities ... Immediately steps must be taken in order to solve the rice drying problem so that not a single per cent of rice crop that is produced is wasted' (Bala n.d. p.3); 'The project is to develop paddy driers for village use ...

these will assist small farmers to save in total a large quantity of paddy each year' (Hurley et al, 1980, p.4-5).

The lobbyists showed a marked degree of indifference to farmers' opinions about loss. One stated that 'the problem of drying rice has been extremely felt (sic) by the researchers and administrators in the field' (Bala n.d. p.2) as though it was not even the farmers' problem he was concerned about. In another case (Hurley 1980, p.3) it is the farmer himself who is the problem because he suffers from a 'lack of appreciation of the losses currently being incurred'! Only one lobbyist (ref.6) used a farmer questionnaire which was of limited value as it consisted of the following five items: (a) Did you meet any loss in boro paddy due to heavy rain in the year (1978)? (b) If yes, please state how much loss occurred in percentage? (c) When did loss occur? (d) Please mention the extent of damage to the quality of the grain (in percentage). (e) Due to the heavy rain there are possibilities of causing damage by the following causes. Please indicate your comment by putting a tick mark: decomposition of paddy ( ); premature germination of paddy ( ); formation of moulds ( ). This survey established through asking 30 farmers these questions that 'seventy per cent of farmers lost the quality of their boro grain to the extent of 51 to 100 per cent' (Karim and Rashid 1979, p.11). Other than this case, lobbyists' recommendations were based upon conventional wisdom, which they reinforced through their own guesses and which provided a basis for their future work.

Lobbyists Rebutted - The Farmer Response To Loss Prevention Technology

At least eight7/ research projects in Bangladesh have attempted to develop loss reducing postharvest techniques. Four of them were projects by the lobbyists. Of these three confined themselves purely to research station assessment of new techniques. Their results on storage are reported in Samajpati et al (1978) and on drying in Sarker (1980) and Bala et al (1980); two recommended their improved techniques and one recommended further research because the actual performance of the drier they built differed from that predicted by theory. In the absence of farm-level experience with these techniques there is little that can be usefully said about their research results. Farm-level research was conducted by the remaining lobbyist (Hurley et al, 1980) and four other studies - which are distinguished from the lobbyists precisely because their authors paid more attention to farm-level experience than conventional wisdom; they are included here only because their original motivation was to evaluate specific loss prevention options. Of these farm-level projects, four (Andersen and Hoberg 1974, Lockwood 1975a, Hurley et al 1980 and Clark and Saha 1980) related to the introduction of drying facilities and one (reported on in Boxall et al 1976) was concerned with improved storage.

The storage research was principally concerned with wheat rather than paddy and particularly with wheat seed storage. No published research reports are available and information about the programme is based upon field visits by the author. The research took place at a time (1975) when wheat was being introduced as a winter crop by the agricultural extension service and shortage of wheat seed had directed attention to the capacity of farmers to preserve wheat seed from harvest to sowing. The project engineer developed an improvement to traditional structures through the use of an insecticidal dust (malathion) and a dehydrating agent (calcium oxide). Farmer response was positive but strongly affected by the heavy promotion - including subsidies - through the extension services. The programme was a fast response by a voluntary field agency (Concern) to a government programme to encourage wheat cultivation. Subsequent surveys (Clay and Shah 1976, Ahmed 1977, and Razzaque 1980) have established that, with a few exceptions, farmers were able to store their seed without outside assistance and the programme has stopped.

The four drying projects were all concerned with the prevention of rice losses in the aus and boro harvests. These studies share three common features. First they were all planned explicitly with the purpose of reducing losses - only one (Hurley et al 1980) actually suggested what loss levels in traditional systems might be. The other three were deliberately cautious in the absence of reliable farm-level measurement of food loss. Secondly, they all addressed in detail the economic and institutional considerations affecting farmer adoption of their techniques. Thirdly, they all concluded that farmers were not very willing to use their technique. Two of them developed solar driers (Lockwood 1975a and Clark and Saha 1980) and the authors in each case concluded that cost and demand factors limited their economic viability at farm-level and sugested they might be most useful in processing fruits and vegetables for dried preparations. The other two tested several different designs of fuel operated driers. The engineers involved were very competent professionals and the design characteristics of their best units were the product of what was almost certainly the highest technical proficiency in any postharvest projects to date in Bangladesh. The reasons why they were not attractive to farmers are complex because the primarily qualitative rather than quantitative nature of drying losses makes interpretation of a farmer's economic assessment difficult.

A discussion of these complexities is reserved for the next chapter in section 4.8 on drying. However, the conclusions of the two projects were clear: 'If continued, the project must shift to the rice traders or mill level at locations in Bangladesh where they are predominant' (Andersen and Hoberg 1974, p.42); and, 'The present attitude is expressed in the statement (by village men) "What a lot of work and trouble to save a little paddy" and this could no doubt be expressed in numerical terms by an economist. But would his wife have the same attitude'. (Hurley 1980, p.29). The reference here to the wife (the farmer's not the economist's) draws attention to the fact that traditional crop drying is predominantly women's work and extension of drying facilities through men possibly constrains their adoption because it is women who have more expertise concerning crop drying requirements and would better appreciate the benefits from efficient drying facilities. Whilst this is clearly insufficient as an explanation of failure it is a most important consideration in evaluating postharvest technical change in Bangladesh and one which, as Hurley points out (pp.28-9) has been neglected.

With this proviso, the unambiguous conclusion of the farm-level experience with supposedly loss-reducing techniques is that farmers are unwilling to adopt them. The resilience in Bangladesh of the high loss lobby is obviously not attributable to the encouragement received from farm-level experience; all too frequently the lobby remains unaffected by such research. But, increasingly, pressure to reform their conventional wisdom is coming from other quarters; the literature reviewed next provides a range of alternative perspectives on postharvest research priorities all of which undermine a single-minded concern with loss prevention.

Some Other Perspectives On The Bangladesh Postharvest System

The development of a broader base to postharvest studies than obsession with food loss and technical change to reduce food loss is reflected in four groups of literature. All of these relate closely to the approach of the present study.

A. THE TASK FORCE ON RICE PROCESSING AND BY-PRODUCT UTILISATION: Under the leadership of the Ministry of Agriculture and at the initiative of the Appropriate Agricultural Technology Cell of the Bangladesh Agricultural Research Council, an inter-ministerial task force was set up in 1977: 1. To provide up to date information on the state of rice processing, and of present and proposed research and development related to rice processing at all levels in Bangladesh. 2. To procure information on the different types of rice processing technologies which may be considered for introduction to Bangladesh. 3. To make recommendations to the Steering Committee8/ regarding the choice of rice processing technology considered for Bangladesh.' (GOPRB 1978b, p.i). The Task Force consisted of seven government nominees and five FAO consultants and after a six week programme of work submitted a report on their findings. Of the seven recommendations two related directly to farm-level postharvest operations; the first recommendation was to undertake 'assessment of rice losses at various stages of postharvest operations' (GOPRB 1978b, p.X). The report placed emphasis upon the problems of policy planning in the absence of reliable loss figures and suggested what needed to be done to improve upon available estimates. The other farm-level related recommendation, the sixth one, was to undertake 'an assessment of drying needs in the country' (GOPRB 1978b, p.xiii). In selecting these two farm-level concerns the Task Force had correctly identified the two topics that were most controversial concerning farm-level postharvest operations. As the literature reviewed earlier has shown the introduction of improved drying facilities was the major area of technical change examined by postharvest technology research; however, the report emphasised the importance of identifying need prior to technological studies and cautiously quoted the experience of one major drying research project (Andersen and Hoberg 1974) which suggested that for farm-level rice drying there was 'insufficient evidence of a general case for modern technology' (GOPRB 1978b, p.56). The report provided clear priorities and, importantly, has provided the justification and support for farm-level loss assessment research.9/

B. SOCIO-ECONOMIC LITERATURE ON TECHNICAL CHANGE: Three authors have been instrumental in advocating more careful consideration of the social and economic factors related to farm-level postharvest technical change - two of whom were members of the Task Force. Harriss has undertaken fieldwork to examine technical change in rice processing technology which is reported in Harriss (1978 and 1979). These articles, which in fact constituted the bulk (63 out of 83 pages) of the Task Force report, dealt respectively with an appraisal of recent rice processing project proposals and with a comparison of farm-level rice processing technologies and custom-milling. This farm-level research was principally concerned with the effects of custom milling on cost reductions and labour displacement in rice processing and provided evidence on both of these. The 1979 paper also drew up a set of nine implications of the pattern of technical change in the rice processing system which related to a broad range of considerations, notably including the dependence of women from poor households upon employment in rice processing. Lockwood, a member of the Task Force appointed by the government, has been involved in Bangladesh postharvest research for ten years and has published extensively on questions related to technology policy (Lockwood 1975 a and b, 1977, 1978, 1979a, b and c, and 1981). These papers, amongst other things, examine the operations of drying, storage and milling and assess the need for and opportunities of technical intervention. They are informed by a detailed understanding of the social organisation of farm-level rice processing and of the institutional barriers to technical change.

Ahmed (1981, 1982 and Ahmed et al 1980) has focused his research on the farm-level need for technical change in rice processing and on the employment implications of the shift from the dheki to the custom mill. His research (listed in Table 3.3) on physical losses up to threshing in the aus season suggests that losses are very low (below one per cent) and that farm-level driers are unlikely to be appropriate. He has also examined the effects of farm-size and tenancy upon losses and concluded that whilst techniques were the same, large farmers were less efficient at preventing losses but that their losses were still very low (1.1 per cent across all operations up to threshing). No differences were observed between tenant and owner-operated farms. He argues that if levels of food loss are properly measured they are shown to be insignificant in determining the rate and direction of technical change and that overall cost reductions, principally through labour displacement, is the chief motivation at farm-level. These findings are developed empirically in an explanation of why farm-level rice drying projects have failed and why rice milling is being affected by technical change with serious consequences for the wage labour opportunities of the poorest women.

The present study is very closely related to the emphases developed by the three authors discussed. It shares the concerns with accurate loss assessment and the economic and social factors relating to technical change in rice processing. Chronologically, the research reported in the next three chapters and Ahmed's research can be seen as responses to the research needs identified by Lockwood and Harriss. In particular, these earlier papers and the Task Force report provide the clearest evidence that the absence of reliable farm-level loss figures is the biggest obstacle to identification of the need for technical change - an absence which the next chapter attempts to respond to.

C. STUDIES ON RURAL WOMEN: A major concern of research on rural women has been their involvement with rice processing. The institution of purdah 10/ restricts women's activities to the farmyard (the bari) where their main farm work is crop, particularly rice, processing. Abdullah (1974) and von Harder (1975) have described these activities and Halpern (1978), Westergaard (1983) and Scott and Carr (1985) have all attempted to evaluate how technical change in rice processing affects rural women. Their analyses, similar to Harriss (1979) and Ahmed (1982) emphasise the effects upon the poorest women of the replacement of dheki technology by custom milling and the need for intervention to develop alternative sources of income. There is a general awareness, reflected in newspaper articles and in conversation with local officials, that some women are adversely affected by the milling changes in technique but the programming needs created by such effects have not been closely addressed by the major rural development programme for women (see Feldman et al 1980); these papers demonstrate the critical importance of correcting this omission. Chapters Six and Seven of this study are concerned to provide a more rigorous empirical base for this and develop the stronger argument that the provision of better income-generating activities for the poorest rural women can provide both the best opportunity to target planned intervention to the poorest rural households and also the best route to improving the social status of women.

D. OTHER POSTHARVEST STUDIES: Finally, there is a heterogeneous group of papers that neither reflect the high food loss, loss reduction emphasis of most research nor the need for accurate loss studies and female employment concerns of the studies reported immediately above. Mannan and Mahmood (1978) and Islam (1980) are both concerned with the energy implications of postharvest techniques, particularly parboiling. Huq and Joarder (1979) and Joarder et al (1980) analyse the incidence of aflatoxins in Bangladesh rice. Nurunnabi and Amanatallah (1980a, b and c) examine the development of rancidity in rice and its relationship to price. Both the aflatoxin and rancidity research was undertaken in collaboration with the project upon which this research was based; they provide evidence (see Section 4.9) that aflatoxins are less severe a problem than some feared but that in certain circumstances they should at least be monitored. Nurunnabi (1975) and Nurunnabi et al (1975) report upon the nutritive changes during different methods of parboiling and varietal performance associated with the different methods. Shahjahan (1975) studied the insect-induced loss, by weight and by proportion of attacked grain, of rice exposed to the Angoumois grain moth Sitotroga Cerealella.

The studies in this final group are all concerned with detailed laboratory experimentation on narrowly defined areas of postharvest grain deterioration. Whilst in some cases an attempt was made to link their findings to policy considerations they are more properly to be seen as basic research necessary to establish the dimensions of a problem prior to detailed applied work. They are sufficiently far removed from applied technical research to leave very broad areas of policy as possible subjects of follow-up research; for example the work by Nurunnabi (1975) on parboiling was directed in major part to the problem of Vitamin B deficiency and the incidence of Beriberi disease. With such research, follow-up studies related to policy can clearly be broad ranging and need not be postharvest specific. They are important studies therefore precisely because they draw attention to the linkages between postharvest issues and other, notably dietary and health, issues of rural development.

3.6 PRODUCT, CAPITAL AND LABOUR USE

In this section the different end uses of rice are discussed in relation to farm-level postharvest operations and the economic valuation of losses. In particular the discussion is concerned with the differences between rice used for on-farm consumption and the marketed surplus. This is followed by a brief outline of the equipment and labour requirements of different postharvest operations. The evidence presented in the tables is from the research project in Bangladesh that the author was involved with from 1978 to 1981. (This project is the main source of data in subsequent chapters; the main features of the project and the organisation of data collection are summarised in the second section of the next chapter.)

The data used in this section are from the preliminary door to door survey on cropping patterns, sources of income and postharvest operations that was undertaken in the eight project villages prior to loss assessment work. The complete enumeration of the eight villages gave a listing of 3094 households. A main purpose of the enumeration was to develop a sample design for the loss assessment work and whilst intensive efforts were made11/ to ensure accuracy it was recognised that collection at one time of information relating to the previous year could not be wholly accurate, even though sufficiently so for sample design. Therefore these preliminary data have not been used to illustrate fine points or address controversial issues but to illustrate fundamental characteristics relating to rice end uses, capital and labour use.

Rice End Uses

Conventionally, postharvest studies employ three main end use descriptions: seed, consumption and market sale. Clearly, there are many others including: kind payment to labour, rental payment, loan or loan repayment, barter and gifts. However all of these eventually become a part of one of the three fundamental farm-level uses of seed, consumption or market sale. In this study it was possible to further narrow the main focus to the differences in farm-level postharvest operations between consumption and market sale. Seed does have special postharvest requirements; the best grains must be selected after threshing and they have to be particularly carefully dried and stored to ensure good germination rates. In Bangladesh this usually involved no special or additional operations but more controlled practice of regular methods used for consumption grain. In this study, germination rates were regularly monitored for seed paddy but otherwise in measuring loss no distinctions were made between seed and consumption uses. There is not perfect substitutability between seed and consumption grain stored at farm-level especially with HYV seed but from a postharvest loss perspective the differences were sufficiently marginal in the case of paddy to ignore the rather complex research requirements to accurately assess the effect, via seed deterioration, of postharvest operations on next season's production. Moreover, whilst farmers had seed-related problems they never expressed the view that these were due to difficulty in preservation.

Conventionally, no distinction is made between consumption paddy and sale paddy when valuing postharvest losses. For sale paddy, quantitative losses are valued at market prices and qualitative losses are valued at the difference between the market price of good quality grain and the price actually obtained. For consumption paddy, imputed values are used, based on these market prices. In the case of quantitative losses this method of valuing lost consumption paddy is perfectly reasonable - the opportunity cost of losses is the imputed value if the farmer and his family simply eat less or the market price of the food bought to replace the lost grain. The conventional method of imputing the value for qualitative losses of consumption paddy is more problematic however.

The conventional approach is justified on the basis that a farm family with a maund of good quality consumption paddy could in fact sell it and buy, say, one and a quarter maunds of poorer quality paddy. Obviously, if the consumption paddy deteriorates in condition this option is not available. The implication of this approach is that a farm family which consumes a maund of rice that has developed a bad smell because of poor drying is worse off than it would have been had the rice not had an off-odour - even if there is absolutely no change in the nutritional value of the rice consumed. There are two problems with this approach.

First, if the farmer knows in advance that his paddy is for own consumption he may choose not to spend labour time on reducing or eliminating qualitative deterioration. Since the harvest time is a busy time and since for most farmers, most of their paddy is produced for own consumption such behaviour is quite plausible. Secondly, the transactions costs associated with paddy sale and purchase may be substantial in relation to the difference in value between good and poorer quality paddy. The smaller is the quantity involved the more likely is this consideration to be crucial.

In fact, both these problems relate to the (incorrect) convention, discussed in 3.2, of valuing losses equally regardless of what postharvest operation they occur in. It was observed in 3.2 that loss of wet, unthreshed grain in field stacks is less costly than loss of dried and threshed grain in the farmyard. Similarly here (problem one) grain which has involved less labour effort in preserving its quality will have a lower value than grain which has been preserved carefully, but more expensively in terms of labour time. And, problem two, grain which is available now in the kitchen for consumption is cheaper than grain which is obtained by going to the market selling some grain and buying back other grain. In both cases, the use of market values to impute the value of losses leads to overestimation of loss.

Now, it is very difficult to quantify these considerations and derive a realistic value for qualitative loss of consumption paddy. However, they suggest the utmost caution in regarding qualitative deterioration of consumption grain as a focus for farm-level loss prevention intervention. This is an important point for recent responses to the mounting evidence of low levels of physical loss have emphasised the significance of qualitative deterioration (SEARCA 1983).

When consumer preferences are sufficiently strong to reject grain for consumption because of quality deterioration it will be appropriate to consider the private economic viability of preventing such deterioration but in Bangladesh the poverty of most producers offers such interventions little prospect now. In the assessment of loss prevention intervention prospects for paddy produced for self-consumption the focus should be very firmly on quantitative losses.

For the marketed surplus, qualitative deterioration that does affect price obviously is included in an estimate of the value of loss. It has been observed12/ that the price differentials due to qualitative differences are of a smaller magnitude in markets where poor consumers are the principal purchasers. Such conditions exist in Bangladesh and therefore reducing qualitative deterioration is less profitable there than for example in a country with an exportable surplus subject to quality-sensitive international prices. However, the extent to which farmers will be affected by greater consumer quality sensitivity reflected in price differentials depends, most fundamentally, on the proportion of their output which is marketed.

Grain is sold as wet paddy immediately after threshing and winnowing, as milled rice ready for cooking and at various other stages between these two. The other two main sales stages are after the raw paddy has been dried and after it has been stored at the farm-level until cash need or favourable prices provoke a sale. Recent aggregate data on the marketed surplus for rice in Bangladesh is not available. Major publications on food policy by IBRD (1979a) and Ahmed (1979) conspicuously avoid strongly endorsing specific estimates of total marketed surplus - though in each case a principal objective was the influence of procurement policies, especially price support operations, on public sector rice procurement. The IBRD study suggested, in an unreferenced footnote (IBRD 1979a, p.18) that: 'Total cash sales on the basis of several small samples appear to be about one-third or more of total rice output. Earlier estimates of 10-20 per cent may have referred only to major wholesale markets'. Ahmed (1979, p.21) cites a 1973-74 Department of Agricultural Marketing Survey of 3000 farmers which showed that one-third of the aman 13/ season rice was marketed. Such estimates refer to the gross marketed surplus, ie total sales by producers without adjustment for grain subsequently purchased by those producers; they offer no guide either to surplus over rural needs or to producers' actual net surplus. They retain some value as a guide to the distribution of postharvest operations after threshing between producers and traders though no precision is possible without information on the distribution of producer sales between raw paddy, parboiled paddy or parboiled rice.14/ Most evidence15/ suggests that the vast majority of producer sales are in the form of raw paddy though petty traders, Barkiwalas and Kutias, purchase small quantities of paddy for farm-level processing and subsequent sale; the size of these operations in aggregate is not known but they occur in all districts and are a main source of income for some landless households and small farmers (Quasem, 1979).16/

Aggregate data on gross marketed surplus is available from earlier agricultural surveys based on large stratified random samples. According to these,17/ growers' sales were approximately 10 per cent of production both in 1963-64 and in 1967-68 with nearly one half of sales occurring in the first month after harvest. With the growth in numbers of farm households it is difficult to reconcile these figures with the IBRD (1979a) belief that cash sales are now about one third of gross output. Unfortunately however, 1977 data from two sources, the land occupancy survey (Jannuzi and Peach 1980) and the agricultural census (GOPRB 1980c) are seriously inconsistent (see Appendix 4.3) regarding growth in the number of farm holdings, owned or operated, so even with reasonable rice production data it is not possible to determine the likely changes in the size of the gross marketed surplus.

Table 3.4

Grain Use Pattern in Four Comilla Villages: 1978-79

all quantities in maunds[a/]

	Quantity	% Total Quantity	Nos. House-holds	Average Quantity (all house-holds	Average Quantity (all house-holds that use)
Production	7521	100	207	36.3	-
Seed	270	3.6	185	1.3	1.5
Consumption	6006	79.8	207	29.0	29.0
Sale	1245	16.6	92	6.0	13.5

Notes: a/ One maund is 37.324 kilograms
Source: IDS Postharvest Project Survey.

Data from the IDS project surveys is not sufficiently complete and is too restricted geographically to provide an authoritative estimate of the size of the marketed surplus but supports the view that the gross surplus is between 10 and 20 per cent rather than one-third of production. There are two types of evidence. The initial door to door survey of 3094 households in the eight villages included questions on the final use of rice from the largest single plot cultivated by each household in 1977-78;[18/] of the nearly 500 tons produced only 11.5 per cent was sold. Secondly, in the four Comilla villages a final use survey of all production in 1978-79 for 207 sample households was conducted. The results are reported in Table 3.4 above. The percentage of grain marketed was 17 per cent and only 44 per cent of producers had any gross marketed surplus - the other households met their cash needs primarily from sale of jute, wheat, oil seeds and winter vegetables. Only 13 of the 207 sample sold more than one-third of their production. Similar results with a total gross marketed surplus of 13 per cent are reported by Ahmed et al (1980) in their study of *aus* postharvest practices in two villages of Mymensingh district. In these villages at least, most producers and the vast majority of production are not directly affected by market determined price penalties for qualitative deterioration. It should be noted though that Comilla is a deficit district with less inequality in land ownership than most districts. In surplus regions and on large farms these price considerations in postharvest operations will obviously be more significant.

A final consideration, that will influence losses of marketed grain but not grain for on-farm consumption, is the timing of marketing. For most small producers in Bangladesh there is no net19/ marketed surplus because needs outweigh production; their sales pattern is largely determined by short-term and often not easily predictable cash requirements and they are not always able to exercise free choice on sales patterns (Quasem 1979). Frequently, their sales occur immediately after threshing and winnowing, and the price they receive is reduced because the grain is still wet. Obvious determinants of a free choice decision on the stage at which the surplus is marketed are price expectations and the risk of storage loss. Farouk (1970) has shown that seasonal price movements in Bangladesh can be related to storage costs including losses though in Bangladesh most of the benefits of seasonal price movements actually accrue to traders (Quasem 1979). It is not proposed to examine this specific issue here but to mention a more general problem of the relationship of marketing practices to the total level of postharvest losses.

Restricting postharvest studies to farm-level losses imposes an artificial boundary on choice of intervention strategies because the strategy to minimise farm-level losses or, more precisely, to maximise net farm income through sales, may not correspond to the maximisation of the net value when the whole processing and distribution system is considered. A simple hypothetical example is where farmers sell wet grain at harvest (because in their assessment it is a better (less risky) economic strategy than realising the higher price of selling dry grain) whereas the highest net value from the whole processing chain would have been achieved if they had dried it because they can do it more cost effectively than the trader or the government procurement centre.

Although, as the data immediately above describe, the marketed surplus is small, it is growing, and increasingly, farm-level practices will be affecting risk of loss further down the processing and distribution system. Moreover, as yield increases occur the risk of loss in existing farm-level postharvest practices will increase and encourage marketing strategies, eg selling wet grain, to restrict the farm-level income costs of losses. The development of market institutions, especially grain quality monitoring, will restrict farmer gains from such practices but will not necessarily lead to greater investment in on-farm processing capability.

To summarise this discussion of end uses of rice there are two significant differences between consumption and sale uses which will affect the valuation of losses. The first is that qualitative deterioration of paddy for own-consumption will often not be a significant consideration influencing farmer adoption of loss-preventing techniques even though the same level of deterioration may result in economic loss when paddy is sold. The second is that the flexibility possible in the timing of marketing will influence farmer adoption of loss-preventing technical change for sale paddy; grain driers are the classic example here and a possible development as the marketed surplus increases is that farmers will reduce the risk of losses by early marketing, rather than invest in driers. However, the gross marketed surplus, as yet, is a small proportion of total output and loss-reducing technical change that is designed to benefit small farmers would have to be directed towards quantitative losses of paddy grown for own consumption - the prospects for this type of technical change are evaluated in the next chapter.

Labour and Capital Use in Postharvest Operations

Tables 3.5 and 3.6 provide details of labour and capital use in postharvest operations. They show that traditional postharvest practices are extremely labour intensive and employ very little capital.

Average labour time per maund of paddy for cutting, stacking, transportation, threshing and winnowing, parboiling, hulling and polishing was 21 hours.20/ This compares with an average labour time per maund in rice production activities up to cutting varying between 2 and 2.5 man-days per maund for most production systems - though in the case of local transplanted aman the labour costs are generally somewhat higher at over three man-days per maund.21/ Differences in sample areas and sizes and data collection methods limit accurate comparisons but very roughly, it can be concluded that the postharvest operations - which include cutting under our definition - require approximately the same amount of labour (family and hired) as for field operations.

So far as hired labour is concerned, a critical issue is the share of their earnings which is derived from postharvest work. The IDS project survey evidence on annual labour earnings showed that 42 per cent of all agricultural labour earnings were derived from postharvest work. As Table 3.5 shows, for men it is the cutting, stacking and transportation work which is their biggest source of postharvest earnings and for women it is _dheki_ work. Table 3.5 also shows that subsequent to cutting and transportation the share of hired labour in total labour drops substantially and that, subsequent to threshing, female labour is almost universally employed.

Table 3.6 gives details on the ownership pattern of the eleven postharvest items most commonly used; there were thirty-eight different items reported altogether but none of the others were used by more than 100 of the 3094 households in the survey. 92.4 per cent of the survey households owned some postharvest equipment. The average cash costs of postharvest equipment including storage structures was just under 92 taka, about six dollars at 1979 exchange rates. The households were asked to provide current costs and if a purchase had not been made in the last year to provide an estimate of current costs. Estimates were required for 34.8 per cent of the cost data provided and for 46 per cent of the 33,436 items recorded no cost was given because no item had been purchased and the respondent was not able to provide an informed estimate of current market cost. This high proportion of 'don't-knows' reflects the fact that many items have a long life-time so purchase is infrequent and that, in some cases, they are home-made (see Table 3.6).

The average household value of postharvest equipment (92 Tk) is slightly misleading because it includes all households who own any postharvest equipment. However, the average value of all the eleven items listed is only Tk 138 and even after allowing for ownership of two or three of certain items such as storage baskets average household investment will only be around Tk 150.

The single most expensive item, the _dheki_, is almost certainly lower than a national survey would indicate because in one of the two survey areas there was easy and cheap access to suitable wood from the nearby jungle.22/ However the chief reason why _dheki_ costs have not resulted in a higher average for total equipment costs is that only 63 per cent of the households reporting ownership of postharvest equipment owned a _dheki_. It was very common for the _dheki_ and other items to be made freely available to households that needed to use but did not own one. It

Table 3.5
Labour Use in Postharvest Operations

Operation a/	Sample Size	Total Quantity (maunds)	Total b/ Time (hours per maund)	% Hired c/ Labour	Labour d/ Type (%)
Cutting, stacking and transportation	1949	16,129	7.8	70.8	99.4 male only
Threshing and winnowing	1451	10,871	3.7	13.9	16.4 male only
Drying after Threshing	1406	1808	(19.5)e/	2.3	98.8 female only
Soaking and parboiling	2033	12,895	2.1	5.1	99.2 female only
Drying after parboiling	2021	12,869	(16.1)e/	4.8	99.0 female only
Dheki Husking polishing and cleaning	1016	4,642	7.9	9.3	99.5 female only

Notes: a/ There was considerable variation in the way that farmers organised their postharvest work. For example: i) cutting and stacking; ii) cutting, stacking and transportation; iii) cutting stacking, transportation and threshing and; iv) cutting, stacking, transportation, threshing and winnowing were the four different groupings in which farmers categorised operations starting with cutting. Therefore, for cutting there were four different estimates of labour time depending on what other operations were also included. The survey results, from 1978-79, were derived from over 2000 rice farmers from a population of 3094 rural households. Each farmer gave information for his single biggest plot in a specified season and all three seasons were included in the sample. Altogether, there were 35 different groupings employed by the farmers to identify postharvest labour use. The six groupings employed in the table together constitute a complete system and were the groupings most frequently used - but, as the sample size and quantity columns indicate, individual farmers included at one stage are not necessarily included at other stages. This is either because they did not follow this particular grouping all the way through or because they sold their crop at an intermediate stage.

b/ Time per maund was calculated on each farm individually and then the average time was derived. This result is sometimes quite different to the division, for a particular operation, of total time by total quantity indicating that there is considerable variability in labour productivity.

c/ Percentage hired labour was not calculated as the average of results for individual farms. The results given are total hired labour hours, including permanent farm servants, as a proportion of total labour hours for all farms.

d/ Unfortunately, gender-specific labour time is not available. This column gives the percentage of occasions that labour was either only male or only female. This approach is useful for all the groupings except threshing and winnowing where (with threshing largely men's work and winnowing women's work) it doesn't tell us very much.

e/ Drying time per maund is not labour time but total hours of drying. It is not possible to derive meaningful estimates of labour time in drying.

Source: IDS Postharvest Project Survey.

Table 3.6
Farm-Level Postharvest Equipment

Item	Nos. House-holds	% Total House-holds	Total Nos. Owned (Tk)	Average Unit Cost	Manu-facturer	Average Life (years)
Sickles	2465	80	3396	3	Artisan	2
Bamboo Fork (for collecting straw)	702	23	1160	3	Self	2
Winnowing Tray	2652	86	3042	4	Artisan	2
Winnowing Sieve	827	27	937	4	Artisan	3
Carrying Basket	967	31	1805	5	Artisan	3
Aluminium parboiling pot	1893	61	2802	40	Factory	6
Wooden Rake (for turning drying paddy)	1621	52	1734	4	Artisan/ Self	4
<u>Dheki</u>	1801	58	1804	55	Artisan	18
Bamboo Storage Baskets:						
Narrow-mouthed	1875	61	3532	3	Artisan	5
Wide-mouthed	1132	37	2959	6	Artisan	3
Large Earthenware Storage Containers	1516	49	2599	11	Artisan	9

Source: IDS Postharvest Project Survey.

should be noted also that none of this equipment is replaced annually so the costs of equipment per unit throughput will be very low in comparison to labour costs. The only items not covered by the survey were the land used (the farmyard) to carry out the operations from threshing onwards and the livestock sometimes used in threshing and, very occasionally, with carts for transportation.

The most critical conclusion to be drawn from these statistics is that, in the absence of high food losses, cost reducing technical change will necessarily be labour saving. If food losses in a particular operation are high it is possible to conceive of a change in technique that would use more labour to reduce losses and result in a net increase in output value. Clearly though, unless there are high food losses, the only substantial way to reduce postharvest unit costs is to save labour. Thus capital investment in postharvest operations, unlike that for production inputs such as fertiliser and irrigation equipment, will not generally involve complementarity with labour resulting in increased labour use, but substitution for labour resulting in decreased labour use. Therefore, careful identification of which labour, family or hired, male or female, is displaced and when it is displaced, slack season or peak season, is crucial. The two types of technical change currently occurring, the introduction of pedal threshers (see Chapter Five) and of mechanical rice mills (see Chapter Six) are both characterised by labour substitution.

NOTES

1. Jannuzi and Peach define the functionally landless as those households owning less than half an acre of land other than homestead land. This category is not entirely satisfactory as a means of addressing inequality in land rights because it takes no account of tenancy; however, their data show that only about a quarter of the 28 per cent entirely landless in 1978 rented in land.

2. In the first Five Year Plan planned allocation in agriculture was 26.3 per cent and in the Two Year Plan (1978-80) it was 27.5 per cent (GOPRB 1980a p XII-6).

3. In fact, the chief competitor with the dheki is the huller mill not the automatic rice mill and it seems likely that the reference here is meant to be to huller mills.

4. For example, a scheme submitted for the plan by the Comilla Cooperative Karkhana (workshop), whilst involving a total of 6.5 crore (65 million) taka for manufacture of pedal threshers, actually required only 36 lakh (3,600,000 Taka) from the government, since a revolving fund was proposed, financed by sales.

5. Salahuddin (1980 p.122) points out that the Plan mentions research on the _dheki_ but no specific proposal has been made.

6. It is relevant, of course, to ask whether a given public outlay would save more food, help poor people's nutrition more or generally show higher returns in non-rice crops. It is possible that a small level of expenditure (before diminishing returns become a constraint) on a technique already available (low R and D costs) for a non-rice crop with high losses would be much preferable to R and D on rice postharvest technology. How small, which technique and what crop are all unknowns that we do not attempt to address here.

7. There have been some other smaller initiatives, apart from these eight projects, in relation to grain storage and drying and mainly by foreign assistance agencies such as the Seventh Day Adventists and UNICEF. Samajapati et al (1978) also list research project reports emanating from the Department of Agricultural Engineering and Basic Engineering, Bangladesh Agricultural University, Mymensingh which include studies on harvesting, storage and drying, published between 1970 and 1974. Lockwood (1979c) also lists (then) current postharvest projects.

8. The Steering Committee was a high level inter-ministerial body set up "to deal with matters relating to policies and long term programmes in the area of postharvest technology" (GOPRB 1978b pXV).

9. One direct result has been to set up, with FAO funding, a farm-level postharvest food loss assessment project; though much delayed this project eventually started in 1983 and a final project report was expected in 1986.

10. _Purdah_ is the Islamic practice of encouraging seclusion of women.

11. Data collection was preceded by two weeks of staff training and by village surveys in the project villages to assist mutual familiarisation of project staff and villagers. The questionnaires were exhaustively dry-tested, printed in the local language and were accompanied by a 70 page instruction manual. The researchers lived in the villages during data collection, there were several cross-checks built in to the questionnaires which were

systematically edited and if necessary, returned for correction/checking.

12. Lockwood (1975b, p.12).

13. _Aman_ accounted for 57 per cent of total rice production in 1973-74 (compiled from GOPRB 1980b, pp.28, 44 and 54).

14. Outside of the Eastern districts of Chittagong, the Chittagong Hill Tracts, Sylhet and Ports of Nookhali there is very little consumption of raw (unparboiled) rice.

15. For example Jabber (1980 p.35), in a survey of farmers selling to procurement centres in the 1978-79 _aman_ season, found that only 7.2 per cent of 3,670 maunds total sales were rice.

16. On field trips in Bogra district for example it was frequently reported to the authors by landless households that buying raw paddy and selling parboiled rice was their major source of income.

17. Government of East Pakistan (1965 p.14) and GOPRB (1972 p.4).

18. It should be noted however that only eight per cent of production included in this survey was from the _boro_ season. This was representative of the relative importance of _boro_ production in our villages but is an under-representation of _boro_ nationally. If a higher proportion of _boro_ production is marketed than production in _aus_ or _aman_ the value of 11.7 per cent is an underestimate. The survey in the eight villages showed that the highest proportion of production marketed was in _aman_ (12.8%) but this result may not hold nationally; the fact that the major growth in rice output is based on expansion of an irrigated HYV _boro_ crop suggests that the contribution of _boro_ to marketed surplus in absolute terms is definitely increasing and, very probably, the proportion of the _boro_ crop marketed is also growing.

19. The net surplus equals sales from own production (gross surplus) less rice subsequently purchased from the market. Quasem (1979, p.66) found that forty-nine per cent of sellers subsequently buy back rice, generally at higher prices. Immediate postharvest sales by small farmers are principally to meet consumption needs (Quasem, 1979, p.67).

20. Drying has been excluded from the total because, in practice, it proved impossible to get remotely accurate figures on labour input into drying by questionnaire methods. It is difficult to give a value for labour time spent bird-scaring and otherwise supervising the drying floor because this activity is usualy done in the farmyard when other tasks are also being performed. Women sometimes

use a long length of stick to scare birds while they are sitting preparing meals etc. The time per maund for spreading the paddy, turning, drying, collecting and cleaning it afterwards, we summise, would not be more than half an hour.

21. GOPRB 1979 vols. I-V.

22. _Dheki_ costs are discussed more extensively in Section 6.5.

4

Bangladesh Farm-Level Food Losses: The Evidence Against the High Loss Lobby

4.1 INTRODUCTION

This chapter presents the results of a three year study1/ on farm-level food losses in Bangladesh. They provide detailed evidence (Sections 4.6 and 4.7) that farmers are efficient in avoiding postharvest losses. The highest losses occur in storage of raw paddy but even there the analysis illustrates the infeasibility of cost-effective farm-level loss prevention intervention. Of necessity, the presentation of these results requires, and begins with, description of field study areas (4.2) and loss assessment methods (4.3 to 4.5). The absence of such descriptions has been a drawback of many 'estimates' of food loss levels.

The high loss lobby in Bangladesh has relied on hunches rather than hard facts to rest its case, principally, upon the alleged need for farm-level crop driers during the wet season. Study results (4.8), including evidence on mycotoxin levels (4.9) in wet season rice, are used in this chapter to demonstrate why farmer estimation of the value of drying losses is very different; sufficiently so to explain the failure of all the drying projects attempted so far in Bangladesh.

Following these results a more tentative argument is developed around the relationship between agricultural growth and postharvest losses. It is suggested (4.10) that yield-augmenting technical change is increasingly responsible for higher risks of farm-level postharvest losses especially in the wet season. The main conclusion for research is the need to integrate postharvest studies in farm management programmes and to focus more specifically on farm-level handling of the marketable surplus.

Finally, the Bangladesh results are compared with some of the more reliable study results from other countries. These show (4.11) that measured losses are typically much lower than the figures usually quoted. Loss prevention opportunities do exist but they must be carefully identified. Low losses can justify low cost interventions especially when the grain saved can prolong the life of grain stores into high-price lean season periods. (Boxall et al 1978, gives a detailed analysis of such a situation for farm-level paddy stores in Andhra Pradesh, South India.) However, a generalised assertion about farm-level postharvest operations of 'high losses justifying high cost intervention' is shown, in the case of rice, to be lacking in empirical foundation.

4.2 THE BANGLADESH POSTHARVEST PROJECT: OBJECTIVES AND AREAS OF STUDY

The study results are from a research project of the Institute of Development Studies and the Bangladesh Council of Scientific and Industrial Research, over three years from July 1978 to June 1981. The team addressed itself to three inter-related tasks:

A. The estimation of food losses in traditional postharvest practices, comparing them to the losses incurred with alternative techniques including mechanical milling and threshing and experimental driers.
B. The evaluation of the income distribution effects of these technical changes.
C. Examining the need for intervention in response to the social and economic implications of the existing pattern of technical change.

This chapter deals with the project evidence on food losses in traditional practices. Field work by three senior researchers and eight postgraduate field staff took place between September 1978 and August 1980. Three laboratory research officers (two entomologists and a biochemist) led by one of the senior researchers analysed grain samples sent in from the project villages. Field work was undertaken in two regions of Bangladesh which were previously selected to give representation to all the major cropping patterns across the *aman* (winter), *boro* (early summer) and *aus* (summer) seasons; at the start of the project, 1978-1979, these seasons accounted respectively for fifty-nine, fifteen and twenty-six per cent of total production (Table 3.1).

The four project villages in Chandina thana, Comilla District included large areas under deepwater broadcast *aman* and mixed broadcast *aman* and *aus* as well as areas that were less prone to flooding where both transplanted *aus* and *aman* were cultivated. In Madhupur thana of Tangail District the project villages included large areas under highland *aus*, lowland *boro* and irrigated HYV *boro* or *aus* followed by transplanted *aman*. These six land use patterns (Figure Four) in rice cultivation provide comprehensive coverage of prevailing patterns. (These rice cropping patterns also involved production of winter vegetables and, very importantly, wheat, during the *rabi* season - if a *boro* crop was not grown - and on some highland in Madhupur pineapples or other annual crops were cultivated.)

Their most critical potential significance for postharvest losses lies in the differing climatic conditions prevailing after harvest. Rainfall, temperature and relative humidity records for the two project areas are given in Appendix 4.1. Because of their proximity, the *boro* and *aus* seasons were treated together in one harvesting season that extends from April through the monsoon to September; the late *boro* is in fact harvested after the early *aus* 2/ and there are essentially similar, though erratic, risks of inclement weather at either end of the harvesting season. The *boro/aus* harvest season is the wet season and the significance of harvest-time rain for both quantitative and qualitative loss is examined in detail in Section 4.8.

A second land-use related factor affecting losses turned out to be varietal differences in straw length. These differences affected the efficiency of cutting particularly and also threshing. Unlike other postharvest stages, in these two operations there were statistically significant differences in measured losses between the *boro/aus* harvest and the *aman* harvest. It is important to recognise, though, that statistically significant differences in loss between seasons may have very limited specific policy implications if the economic value of preventable food losses is small in relation to the productivity and distributional implications of loss-prevention interventions.

A third influence of land-use patterns on losses is their effect on the timing of postharvest operations in relation to other demands on available labour and draught power. Specifically, delays in completing postharvest

Calendar of the Six Rice Cropping Patterns in the Project Villages

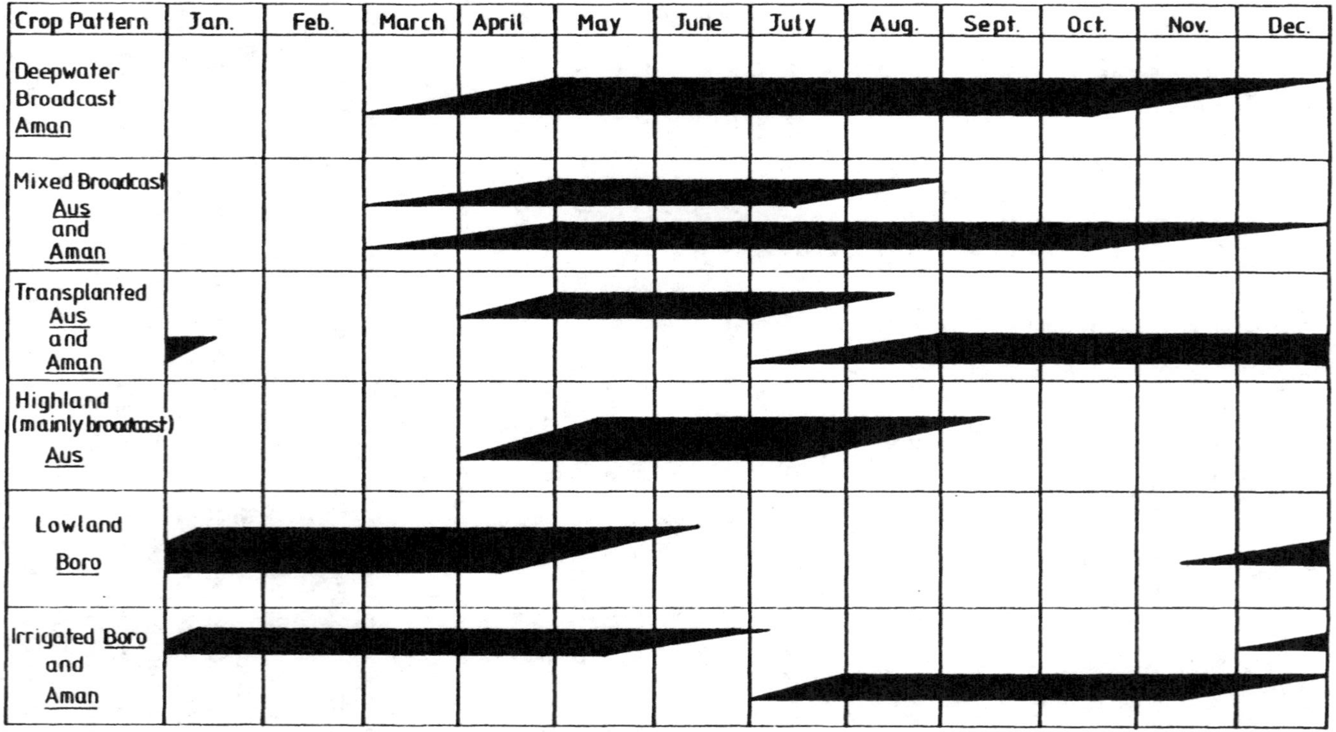

Figure Four

operations because of labour or draught power constraints can increase the risk of losses, especially qualitative deterioration. It is argued below that such delays are becoming more likely as a consequence of the increasing share of HYV boro production in total rice output and that this development provides the most important focus for future research.

4.3 GROUPING OPERATIONS BY LOSS ASSESSMENT METHOD: DIFFERENT CAUSES REQUIRE DIFFERENT METHODS

In chapter three, physical loss estimates were identified as having three essential requirements; that they were accurately measurable, soundly defined and could be expressed in terms of an economic value. Chapter two attributed some of the culpability for inadequate evidence on the importance or otherwise of food loss, to the paucity of studies that had used, and reported details of, reliable loss assessment methods. It is obligatory therefore to describe here, very precisely, the loss assessment methods employed in the study.

Chapter three described the eight postharvest operations for rice in Bangladesh. For loss assessment purposes the two parts of the first operation, cutting and field stacking, are treated separately; for reasons which will become apparent these nine operations are divided into two groups. Group one consists of cutting, field stacking, transportation from field to (farm) threshing yard, threshing and winnowing, storage and milling. The commonality of this group lies in the fact that direct physical losses during the operation itself constitute the major arena for the analysis of food loss.

Group two operations are farm yard stacking, soaking and parboiling and drying. In these operations direct physical losses during the operation itself are usually negligible and the main risk lies in indirect physical loss which is defined as reduction of milling yield following qualitative deterioration. In the case of farmyard stacking, extension of the stacking period when threshing is delayed is the potential source of qualitative deterioration because the grain stack is usually hot[3] and wet. In soaking and parboiling, the risk of deterioration lies in the extension of the soaking period over several days because weather conditions are not suitable for parboiling and drying; parboiled paddy of course yields higher milling outturns than raw paddy and parboiling per se is not a cause of qualitative deterioration leading to reduced milling yield. Drying is associated with qualitative deterioration leading to reduced milling yields because of delays in drying, insufficient drying, and too rapid drying. Very occasionally, prolonged rain prevents drying for so long that the level of deterioration forces complete discard of the grain. As the results in 4.8 show, this occurence is extremely uncommon.

This second group of operations are usually considered to be particularly problematic in the wet season and it is in this context that they are evaluated. The logic of the grouping is in part dependent upon the fact that farmyard stacking, soaking and drying are all operations where variation in practice is associated with farmer response to unfavourable weather (rain). In addition to suffering indirect physical loss due to reduced milling yield these operations also result in a reduction in consumer observable quality characteristics.

This grouping into operations which suffer direct and those that suffer indirect physical loss is not offered as a rigorous framework incorporating all sources of physical loss but those which are accurately measurable. The second group of operations are all also quite clearly potential sufferers of direct physical loss through spillage and scattering. Most people who have witnessed these operations in Bangladesh would agree such losses are negligible but the grouping employed is based upon the strong assertion that they are immeasurably small.

These stacking, soaking and parboiling and drying operations, typically, are all farmyard operations where family farm women take a great deal of care in cleaning equipment and sweeping the yard to prevent spillage and scattering. The only method (see next section) to measure loss is through repeating these women's work to collect what they have missed; attempts to do this were always anticipated and produced much amusement and no grain. With the farmyard and equipment often being used for several lots of paddy, sweepings from one lot could not usually be separately identified anyway. For these reasons it was decided not to attempt a rigorous quantification of physical losses due to scattering during farmyard stacking, soaking, parboiling and drying. The best is often the enemy of the good, and the grouping employed here reflects the severely practical considerations of applied analysis.

4.4 CROP DRYING AS A SPECIAL CASE?

An arguable exception to this, for two reasons, is physical loss during drying. First, it appears that a loss assessment method is available; the grain can be weighed and the moisture content taken before and after drying. The difference in dry weights ought to measure loss. In practice, however, the variability of moisture content, both pre- and post-drying but especially pre-, and the inaccuracy of the electrical resistance field moisture meters at high moisture content, cause severe inaccuracy. And this inaccuracy occurs precisely where greater accuracy is required, for an error of say, four per cent on a moisture reading when the true grain moisture content is twenty five per cent will give an error on the initial dry weight of one per cent; a larger error than the per cent loss even careless drying would be likely to produce. Transferring small samples in air-tight containers to the laboratory where more accurate moisture measurement techniques can be used is a possible solution. This has been employed usefully to measure drying rates, but attempts to follow this procedure both in this and other loss assessment projects have not produced accurate results.

The estimate of drying loss through such methods is in fact a measure of the difference between weight loss attributable to drying and the total weight loss. The method presumes that the measurements can be so accurate that any difference in the dry weights can be regarded as physical losses during drying. In other words, it involves measuring the loss of water during the drying process and then subtracting that loss from the total change in weight of dry matter to obtain the loss of food. In theory if there are no losses the estimated dry weight, using the moisture content readings to adjust the estimates of wet weight, should be the same before and after drying. Whilst the drying rate (loss of water) estimate taken alone, should always produce a plausible figure, and inaccuracy will anyway not be obvious, combining this estimate with that of total weight loss in a formula to obtain a drying loss estimate can easily produce impossible estimates when either of the two moisture content and weight measurements are inaccurate. Recent, carefully conducted field work (reviewed in Greeley 1984) has produced negative estimates and impossibly large (20 per cent) estimates more often than it has produced plausible ones.4/ (The problem is analogous to that of independently measuring the potential yield and actual yield for losses upto threshing which is described below.)

Secondly, because of roadside drying, because of poorly prepared drying floors and because of pilferage especially by chickens, crows and sparrows, it can be argued that physical losses are not so small, even if they are immeasurable. Urban based and tarmac-biased postharvest voyeurs, who witness roadside drying exposed to traffic, frequently dispute the assertion that scattering losses are low, but most farms in most villages do not have access to tarmac roads. Drying is most commonly conducted within the farmyard on prepared drying yards. For the same reasons as given below, in the discussion on threshing methods, physical losses in these yards are immeasurably small.

In the case of losses due to birds it has to be conceded that they occur frequently enough even though careful supervision of drying can prevent them. Whilst the measurement problems described above apply with equal force it is possible anyway to demonstrate that these losses are normally insignificant. The yield from an acre of HYV paddy in the project villages averaged 50 maunds or 1.9 tons; to lose even one half of one per cent of this would require birds to eat between 400,000 and 500,000 grains. (Laboratory analysis of storage samples involved frequent weighing of 1000 grains and those weights typically varied between twenty and twenty-five grams.) The number and voracity of birds required to achieve such a loss in a day of drying in the farmyard is left to the reader to determine! Moreover, whilst Addison's sentiments5/ on birds and their food habits may not be universally shared, it is not uncommon for farmers to be unperturbed by their poorly fed poultry adopting self-help methods to improve their diet.

To conclude this discussion on drying losses, it is only fair to emphasise that whilst the introduction of farm level crop driers has been the clarion call of the interventionist lobby it has always been argued for principally on the grounds of qualitative deterioration because of delayed or inadequate sun drying leading to reduced milling yields or loss in consumer desirable quality characteristics and not because of physical losses in drying itself.

Qualitative deterioration in all of the group two operations, was examined over two wet seasons harvests (1979 and 1980). Though there are two types of qualitative deterioration that have been identified - that resulting in indirect physical losses and that resulting in loss of consumer observable quality characteristics - the research undertaken was focused only on the first, physical losses. The reasons for this emphasis in the work and the results of that work are discussed below in Section 4.8 on wet season losses.

4.5 THE PROBLEMS OF DEVELOPING ACCURATE APPLIED METHODS

Returning to group one operations, we divide loss estimation methods further according to the three distinct approaches employed for: a) operations upto threshing; b) storage; and c) milling. The discussion of these approaches is concerned both to give clear definitions and, where necessary, explicit justification of the methods employed and to elaborate, through selective illustrations, the pitfalls to which loss assessment methods are susceptible. The major illustration deals with the problems of measuring cutting losses because unlike, say, storage6/ they have not been widely discussed and yet cutting losses are often thought to be large.

a) Operations Upto Threshing

Physical losses in the operations upto threshing, are the differences between the potential yield of the standing crop and the actual yield obtained by a farmer after winnowing the threshed crop. The per cent loss is the difference between potential yield and actual yield expressed as a percentage of potential yield. Measurement of the actual yield simply involves accurate weighing of the farmers' grain, but there are two ways in which the potential yield can be determined for any particular plot:

1. By performing all the operations from a sample area of the plot, under careful supervision to avoid any shattering loss.
2. By collecting all the grain lost at each stage during the farmer's own operations and adding these grains to his actual yield.

Both these methods require standardisation of moisture content each time the grain is weighed and standardised methods of winnowing and cleaning to ensure that at each point of measurement the weight is a uniform proxy for the food value of the grain. If these conditions are not accurately observed, differences may exist in the milling yield of whole rice and brokens which will affect the accuracy of the measurement of loss.

There has been little experimental work undertaken to estimate losses during these early stages of processing and no definitive study was available to guide a selection between these two methods. Therefore, one of the early concerns of the research was to establish the relative efficiency of each method. The results of this comparison, which are briefly discussed, showed fairly conclusively the enormous practical difficulties of obtaining accurate estimates from method one.

The first method is essentially similar to crop cut methods for estimating yields used in variety trials and in national production statistics; (these have changed little since their first systematic development in the 1920s in Orissa, India, by Hubback as is evident in comparing suggested methods in Yates, (1946) and FAO (1981b). In simple measurement of acre yields, there are two sources of non-sampling error; inaccuracies in measurement methods and shattering losses incurred during the cutting operation.

The comparison of potential and actual yield is designed to measure the shattering losses but experimental design has to confront the problem of distinguishing between genuine shattering losses and apparent shattering losses which are actually inaccuracies in measurement methods. As described in the following paragraph, inaccurate measurement is difficult to avoid and tends to increase the potential yield estimate above its real value.

High <u>interplot</u> variation prevents use of comparative yield estimates (potential and actual) from different plots, even within the same field, and therefore the potential yield estimates must be based on a sample crop cut from the plot being used to estimate the farmer's yield. Randomising selection of sample cuts and choosing small plots of visibly uniform yield - avoiding edge effects, soil fertility gradients and localised pest damage - reduce the possibilities of <u>intraplot</u> variation invalidating or reducing the significance of results. However, measurement of area in these small crop cut estimation procedures introduces another form of bias leading to overestimates of yield, viz., the inverse relationship between the size of crop cut and the degree of error in estimation. The problem is that the smaller the area the larger the edges are in relation to the total size of cut, and therefore any inaccuracy in measurement will cause a higher degree of error in smaller crop cuts; yet to obtain a series of random sample cuts from a small total plot area requires use of a small sample cut. This is a very serious problem; data from the Indian Statistical Institute (Mahalanobis, 1961) show that using one square foot samples gives an estimate of per acre yield four times greater than an estimate obtained for the same plot using a sample of two hundred and fifty square feet. The degree of accuracy depended on the training of field staff, but even with well qualified staff, major problems with the measuring instrument and in achieving randomness were present.

IDS-BCSIR project results, using both rigid frames and ropes and pegs for measuring area, were based on sample cuts, from $108m^2$ plots, of $10 \times 1m^2$, $1 \times 10m^2$ and $5 \times 2m^2$. These gave figures of potential yield averaging ten, thirteen and fourteen per cent greater than actual yield in three different seasons but with a range from minus ten to plus thirty per cent. As the results reported below show, these average figures grossly overestimated actual loss. They confirm the results of Mahalanobis and indicate a marked overestimation bias in conventional small scale crop cut methods. Further refinement of the method to use 'hill'7/ yields rather than area yields provided no improvement. We concluded that high soil fertility gradients made this method impracticable. The method was anyway identified as inefficient on statistical grounds because experiments with it provided a coefficient of variation (CV) of 87 per cent whereas the CV for the second method, described below, varied between 17 and 24 per cent. A comparison such as this is a standard technique (see for example Snedecor and Cochran 1980 p.37) for correct selection of field research methods.

A degree of error in estimating yields of ten per cent may be tolerable in indicative crop cut surveys, but measuring losses, which may be as little as one per cent, obviously cannot be successfully accomplished without a more accurate measure. Crop cut surveys comparing varieties can also ignore this inaccuracy if it is a systematic bias, but to fulfill our objectives the more time-consuming and expensive estimation procedures based on collection of physical losses had to be adopted.

In this second method each operation was independently evaluated by physical collection of the grain lost. During cutting, the farmer's actual yield was based on a sample plot of one hundred and eight square metres (a convenient size for accurate measurement, and large enough to avoid any small cut biases, as occurred in method one). Using three tape measures, two sides and a diagonal are set by a triangle with sides in the ratio 9-12-15; maintaining the diagonal tape, two tapes can be shifted through ninety degrees to meet and give the fourth corner of an exact rectangle (9 x 12m). The corners are marked by long wooden pegs and a rope is run tightly around them whilst the tapes are still in place. Before cutting, the edges are checked and any heads close to or on the boundary rope are cut and placed with the appropriate yield, either within or without the sample plot according to their exact position.

The minimum size of plot was determined by the imperative need to avoid interference with usual farmer practice, as might happen if plot size were too small; to reduce <u>intraplot</u> variability this size was kept to the minimum consistent with that need.

After cutting, five separate areas of two square metres were marked out within the plot, using iron frames, and very carefully gleaned; each area took approximately fifteen minutes for two people to glean. After gleaning this grain was weighed and adjusted for moisture content, which in all estimates was standardised at fourteen percent. After threshing, the farmers' actual yield was weighed using a spring balance and adjusted for moisture content and for losses during field stacking and transportation and threshing that were estimated separately (see below). These two yields when standardised for area and combined provide the potential yield; the standardised gleaned grain yield expressed as a percentage of potential yield gave the estimate of loss.

Assuming that the areas are measured accurately the biggest drawback of this method is the difficulty of gleaning in a muddy field; a gleaning estimate is not possible where there is standing water or a soft muddy field, because fallen grains are difficult to see and pick up. For this reason a number of sample plots did not yield results for cutting losses. However, there is no reason for believing that shattering losses in muddy fields would be higher than average and quite possibly they might be lower.<u>8</u>/ This omission is not problematic therefore.

Losses during <u>field stacking</u> and <u>transportation</u> were estimated in a similar manner by collecting all scattered grains. In the case of field stacking this was done by placing a tarpaulin underneath each field stack; once the field stack was removed for transportation to the farmyard all the fallen grains were collected. During transportation two, and sometimes three, other men walked beside the labourer carrying the grain by head or shoulder-pole and they caught in baskets or picked up any grain that fell. (This is an arduous task!). The loss estimates were then based on the standardised yields of the gleaned and fallen grains as a percentage of the potential yield. In the case of stacking the potential yield included stacking, transportation and threshing losses, but in the case of transportation the stacking losses have already occurred and are therefore not part of the potential yield. Ten results for each type of loss were completed in one day's work by two field staff.

The estimation of threshing loss was obtained by re-threshing a known percentage of the farmers' straw (often from the same sample plots as used in the cutting loss estimate). Physical losses can also occur through grain lost in muddy threshing floors. A reasonable method for measuring such loss is not available; however such losses could and invariably were avoided by care in preparing the threshing floor. Sloping the floor and compacting the earth, drying the floor with husk, covering it with bamboo or extending the pre-threshing stacking period and allowing the floor to dry out were all techniques sometimes employed to deal with the problem. One or more of these practices was almost always necessary during the wet season harvest. Of course, there are cases of carelessness or where a muddy threshing floor has had to be used because of the constraints affecting the availability of labour power and bullock power for threshing. However, such losses are immeasurable and under usual farm management practice are of no significance. Similarly, winnowing is not a measurable cause of loss due to scattering; it is a farmyard operation and, as with scattering during the group two operations (farmyard stacking, soaking and parboiling and drying), the care usually taken by farm women prevents any meaningful estimate of what can only be a negligible loss.

Threshing can also be the cause of reduced milling yield i.e. an indirect loss as in group two operations. This is because the grain is subject to stress at the time of separation from the straw and this can lead to internal cracking resulting in an increased percentage of brokens during milling. However, comparing laboratory milling results from samples threshed by different traditional methods and by pedal threshing there was no statistically significant difference in the percentage of broken grains. It seemed reasonable therefore to base the loss estimate only upon grain remaining unthreshed by re-threshing the straw.

The re-threshing involved careful hand-stripping and then winnowing in a manner identical to the farmers' winnowing, to avoid green or hollow grains that they would normally separate. The estimate of loss was then based on this yield, standardised, as a percentage of potential yield; potential yield in this case is the actual threshing yield plus the re-threshing yield and excludes grain lost at earlier stages.

b) Storage

The estimation of storage losses is more protracted. It requires weighing of grain into store and each time grain is removed from store over what is often a period of several months before the store is exhausted. A sample of grain is taken at each time of weighing for laboratory analysis to determine the extent of insect damage. Additionally, account has to be taken of different commodities (raw paddy, parboiled paddy, parboiled rice) and store types. Technical details of these requirements as well as the loss assessment methods given below have been described in Huq (1980) and Huq and Greeley (1980). (These papers also provide details of interest to storage entomologists regarding the moisture content and insect infestation patterns.) Immediately after the harvest, sample stores were identified and selected. Once a store had been selected detailed information on construction, capacity, price, age, location and physical defects were noted and a brief history of the stored grain from sowing was also recorded. The weight of the grain in each selected store was taken using scales marked accurately to 200 grams, and a representative sample was drawn from the store to provide baseline information for loss assessment. A store record card was attached with each selected store to avoid mistakes in identifying the store in subsequent visits (a farmer may have five or six practically identical stores). At each removal, the date and quantity of grain removed were noted and moisture content and temperature determined.

The estimates of paddy/rice loss are based on physical dry weight loss; i.e. weight loss measured, as just described, by accurately weighing grain in and out of store and standardising moisture content at zero. The figure so obtained for total weight loss consists of two separate types of loss: reduction in weight due to insect (and mould) activity and pure "theft" usually by rodents. Insect losses can be estimated in a variety of ways (see below) but there is no reliable method for estimating rodent losses. The usual procedure is to estimate insect losses, subtract them from total physical weight loss, and ascribe the remainder to rodents. This was the method adopted here after estimating per cent insect losses by the laboratory analysis of samples.

Samples were taken at the time of removal by picking up handfuls from grain being removed until a one kilogram sample was collected. Use of grain removed and evidence of rodent damage and insect species present were recorded on the sampling record form. There are two commonly used methods (Harris and Lindbad 1978) of estimating insect losses in grains. An accurate estimate of weight loss can be obtained by comparing the dry weight of a standard volume of sieved grain from the sample with the dry weight of a standard volume of a baseline sample; the percentage weight loss in a sample can then be determined based on the loss in weight. To estimate the total insect loss in store this sample loss figure is multiplied by the proportion of total grain removed; summing over all removals given total store loss due to insects.

The other method for estimation of weight loss is by the count and weigh method. 1000 sample grains are divided into undamaged and damaged fractions; the grains in each fraction are counted and weighed; and the percentage loss is calculated by a formula that measures weight loss of damaged grains and expresses it as a percentage of the weight of 1000 sound grains.

The disadvantage of this method is that hidden infestation results in an underestimation of loss. Before the emergence of the adult insect, earlier stages of development are responsible for loss of weight inside the grain but because grains are visibly undamaged (no exit hole) they are counted amongst the sound grains. At low and high levels of damage this method and other methods using similar principles are not dependable (but see Proctor and Rowley 1983 for a refinement that should improve utility.)

The laboratory analysis also included estimation of hidden infestation and a standard milling test on each sample to help assess the methods described above. As anticipated, the standard volume weight method gave the more accurate results and it is these which are reported below.

c) Milling

The final operation in group one is milling (husking and polishing). Losses in this operation are not based upon absolute physical loss as in earlier operations but upon the difference between obtained milling yields and an optimum milling yield derived from laboratory estimation.

This comparison has to be based upon two samples from the same lot of grain to avoid differences arising due to differences in the pre-milling grain quality.

This approach is complicated by the need to set a desired standard for polishing; highly polished rice will have less bran and therefore weigh less than underpolished rice. Sophisticated measures have been developed to meet this need (IRRI 1978) but the basic requirement can be met by securing a standard level of polish on the laboratory milling equipment that approximately reflects average levels of polish for the consumer group concerned.

Further complications lie in the estimation of percentage broken grains. Brokens are rarely a substantial cause of complete physical loss because in Bangladesh they are invariably used, but not always in the same manner as whole rice. This introduces a valuation problem, and even when they are used with whole rice their food value is reduced because of greater physical loss of solids during cooking. These issues are taken up in more detail in chapter six but the presentation of results below includes the estimation of both physical reductions of milling yield and the percentage of broken grains. Milling yield is the weight of rice including brokens as a percentage of the initial paddy weight at a standard 14 percent moisture content. Broken grains are the weight of brokens as a percentage of the total rice yield.

To summarise, six operations have been incorporated in the study of direct physical losses - cutting, field stacking, transportation, threshing and winnowing, storage and milling. For the three other operations, farmyard stacking, soaking and parboiling and drying, the estimate of loss is based only upon consequent reductions in milling yield. Further, for this second group, the risk of loss is analysed specifically in the context of the wet season (aus and boro harvests) because of the greater risks of these losses occurring then. Although fieldwork to collect data on these sources of loss was conducted over two years in the two project areas the sample sizes, except for storage, were not very large because of the time required for making each estimate. As evidence of this, the instructions to field staff, including a completed example, for a set of cutting to threshing experiments are provided in Appendix 4.2. These forms were used for the transplanted aman crop in 1979 - 1980. They have been selected because they include details of the methods employed to test the two differing approaches in estimating cutting to threshing losses described earlier. They provide four separate composite estimates of cutting to threshing loss; an independent estimate of cutting loss and an independent estimate of threshing loss for each plot. (Measurement of field stacking and transportation losses were undertaken in separate experiments). Their inclusion here seeks to underline the fact that whilst a single number giving the overall loss estimate is all that most non-specialists are concerned with, the volume of effort required to produce such a figure is substantial. Collectively, these notes on applied loss assessment methods help to explain the bias of funding bodies towards intervention rather than measurement of loss. The next section, on the fruits of these methods, shows why this bias was harmful.

4.6 ESTIMATED FOOD LOSSES AND ECONOMIC IMPLICATIONS

The best estimate of average food losses in farm-level postharvest operations on paddy for the eight project villages is 6.9 per cent. The breakdown of losses by operation is given in Table 4.1 and the notes there explain the various weighting procedures involved to derive the single figure for the whole system. No losses for milling are included because, as Table 4.2 shows, traditional milling of parboiled paddy yields as much rice as modern milling.9/ The breakdown of pre-storage and storage losses by season and by crop type are given in Table 4.3.

Obviously, an increase of seven percent in domestic rice availability through loss prevention would be most welcome. The problem lies in realising such benefits cost effectively when this low total loss is spread across five operations.

The operation with the highest loss was storage. Amongst the three stored commodities, raw paddy suffered the highest average loss of 3.3 percent. This is a consequence of the longer storage period; both the differences in losses and in storage periods are statistically significant. Within raw paddy storage, differences in losses between each pair of store types were not statistically significant; the highest losses were in bamboo type structures at 3.6 per cent (Table 4.4).

This loss figure from the most frequently used store type corresponds almost exactly to the upper boundary of the (95 per cent) confidence interval for average loss of raw paddy (3.3% + or - 0.3) and is therefore a convenient value to use for a consideration of the requirements for cost-effective intervention.

Table 4.1
Measured Food Losses in Rice Postharvest Operations a/ (1978-1980)

Pre-storage Operationsb/	Per Cent Loss	Sample Size
Cutting	1.46	55
Field stacking	0.50	100
Transportation	0.53	100
Threshing & winnowing	1.79	98
Total (weighted)c/	4.2	
Storage (av. months)		
Raw paddy (4.4)	3.3	241
Parboiled paddy (2.4)	2.4	46
Parboiled rice (3.0)	2.5	62
Average (weighted)d/	2.8	
Total (weighted)	6.9	

Notes: a/ The estimates for each season and their standard errors are in table 4.3. Details on the raw paddy storage estimates are in table 4.4.

b/ Loss estimates for broadcast and transplanted crops were obtained in each season and the weights used to calculate these overall loss estimates were based upon the contribution of each season and each sowing method to total rice production using the (then) current production statistics compiled by the Bangladesh Bureau of Statistics. The weights used were: Aman broadcast 0.13; transplanted 0.45; Boro and Aus broadcast 0.23; transplanted 0.19.

c/ Losses are not additive since all the percentages have different bases. In this case the difference is small. From a hundred maunds of paddy a farmer loses 7 maunds according to a straight aggregation though the per cent loss column of the table and the corrected weight loss is 6.9. With national rice production in 1982-83 of around fourteen million tons, this loss, at a national level, equals more than 950,000 tons of rice.

d/ The weights used were 0.19 raw paddy; 0.39 parboiled paddy and 0.41 parboiled rice. These weights are derived from the storage pattern of over 17,000 maunds of paddy in the eight project villages from the Aman 1977-78 and Boro and Aus 1978 harvests. The weighting is based on the storage pattern of the major variety cultivated by each of 2,148 households. When considering losses from total production, as opposed to losses from total quantity stored at farm-level, different weighting is required. There has to be a downward adjustment of total weighting to allow for that proportion of paddy sold at harvest and an upward adjustment to allow for storage of the same quantity in different forms. The proportion marketed immediately was at least fifty per cent of the gross marketed surplus (11.5 per cent in our villages). Since this percentage (5.8) was only slightly greater than the percentage of the total production that was double stored neither adjustment was made. This leaves a negligibly small bias towards overestimation when the percentage loss is applied to production rather than storage statistics.

The sample mix by commodity is heavily biased towards raw paddy. The principal reason for this is the fact that the initial perception of the research team, confirmed by the experience of the two years fieldwork, was that if opportunities to prevent food loss during storage did exist they would be most likely in raw paddy storage. Raw paddy stores suffered the highest losses even though the main store types were the same for all commodities.

In paddy stores insects feed more on the rice than on the husk which means that weight loss of paddy underestimates food loss of rice. If the loss was evenly distributed between rice and husk, the proportion of loss from rice would be 72 per cent (the milling yield). In fact, preferential insect feeding results in about 90 per cent of the total weight loss coming from the rice and only 10 per cent from the husk.

To allow for this a conversion factor of 1.25 (90/72) was applied to the proportion (50 per cent in fact) of raw and parboiled paddy lost because of insect attack.

Source: Tables 4.1 to 4.10 are all based on the fieldwork of the IDS postharvest project. All have been previously published: tables 4.1 to 4.3 in Greeley (1982a); table 4.4 in Huq (1980); and tables 4.5 to 4.10 in Greeley and Rahman (1981)

Table 4.2
Losses During Rice Milling

Milling method	Per cent milling yield	Per cent broken grains	Sample size
Dheki	72.02 (0.46)a/	28.28 (4.50)	20
Engleberg-type huller rice mill	69.94 (0.33)	30.54 (4.52)	20
Laboratory modelb/ modern milling	71.87 (0.28)	8.89 (1.91)	20

Notes: a/ Figures in brackets are standard errors.
b/ The samples were husked in a laboratory rubber-roll Satake Testing Husker, model THU-35A manufactured by Satake (Japan) and polished in a truncated stone cone laboratory rice mill manufactured by Olmia (Italy).

Table 4.3
Postharvest Operations: Physical Food Loss by Season and Crop Type

	Aman						Aus/boro		
Pre-storage losses	Broadcast			Transplanted			Transplanted		
	% loss	Sample size	Standard error	% loss	Sample size	Standard error	% loss	Sample size	Standard error
Cutting	2.98	17	0.71	2.06	18	0.35	0.34	20	0.08
Field stacking	naa/	na	na	0.53	50	0.04	0.46	50	0.03
Transportation	naa/	na	na	0.67	50	0.03	0.34	50	0.02
Threshing and winnowing	1.44	20	0.18	1.45	20	0.28	0.60b/	38	0.08
							3.64	20	0.42
Storage losses	Aman			Aus			Boro		
	% loss	Sample size	Standard error	% loss	Sample size	Standard error	% loss	Sample size	Standard error
Raw paddy	3.3	173	0.20	3.2	40	0.39	3.5	28	0.71
Paddy parboiled	2.7	18	0.80	1.9	12	0.43	2.2	16	0.54
Rice parboiled	2	55	0.17	3.1	7	0.66	-	-	-

Notes: a/ Field stacking and transportation were not measured for the *aman* broadcast crop because field conditions prevented accurate or meaningful measures. Frequently, in the deepwater fields, water was still in the field and the paddy was stacked straightaway on a small boat for transportation home. Where boats were used losses for these two operations would have been lower than for a transplanted crop. Where boats were not used, losses would have been similar to the transplanted *aman* crop. (Scattering losses during stacking would be higher in all probability but shattering losses lower in both operations).

b/ 0.60 per cent for a transplanted crop and 3.64 per cent for a broadcast crop: the other pre-storage operations are based on transplanted crops as the broadcast crop is sown mixed with *aman* and harvested while the *aman* is still standing in the field which prevents measurement of harvesting losses. The apparently high broadcast crop estimate is almost certainly greater than the true average since the twenty samples were purposively selected to give a range of pre-threshing stacking periods. They have been included in the average results since their low weight in the total means that they will not affect average loss levels significantly. However, they have not been used for the tests on the effects of straw length on threshing efficiency.

It is worth reiterating first that actual losses on any one randomly selected farm will be lower than this value in more than fifty per cent of selections because the distribution is skewed to the right.10/ Also, the total period of storage (4.4 months) from the sample was slightly longer than the average period from the overall village survey results for raw paddy (3.98 months). For both these reasons the estimate of the economic benefits from reducing losses overestimates the potential benefits for most farms. It should be noted though that there are some areas in Bangladesh where only one paddy crop is grown and where the storage period therefore tends to be longer (for those farms with sufficient production to maintain consumption stocks). The only substantial areas of Bangladesh which are still mono-crop are the low-lying areas of Sylhet (the haor areas) where a boro crop is produced. Our results are likely to underestimate losses in store in these areas.

Table 4.4

The Level of Losses for Raw Paddy in Different Store Types (all seasons)

Store type	No. of stores	Av. period in months	Percentage loss by weight Total loss	Insect loss	Rodent loss
Bamboo structure	112	4.38	3.56	1.77	1.79
Earthenware	43	4.33	3.00	1.79	1.21
Bag	76	4.42	3.25	1.48	1.77
Others	10	4.42	2.95	1.29	1.66

Note: The distribution by store type in the sample broadly reflects the population distribution in the sample villages (Huq and Greeley 1980). However, differences in losses between store types were not statistically significant. The 'average' period is defined here, following convention, as the average of the total life of the stores; assuming a regular and even removal pattern, the true average period is of course only half of this conventional definition.

Losses in bamboo stores were caused almost evenly by insects and by rodents. Loss prevention would therefore require introduction of new techniques or practices that would prevent access to rodents and the development of insect infestation. The physical characteristics of the structure do not allow improvements that would obtain these objectives. Unlike in South India and the Dry Zone of Sri Lanka, the rainfall levels are high enough to prevent use of structures outside of the house;11/ therefore the relatively simple and cheap rodent-proofing improvement (Raman et al 1976) developed for outdoor bamboo structures cannot be used in Bangladesh.

These bamboo stores are kept inside in Bangladesh and, of special importance, in the sleeping areas of farm homes which means that the use of fumigants to prevent insect infestation is only possible where the air-tightness of the storage structure is absolute;12/ in practice this condition is never met and to introduce mud-coating methods that would meet it is likely to be a difficult task because there is no tradition of mud-coating the top of the bamboo basket. Alternatives to fumigation such as chemical insecticides or traditional insecticidal agents like dried neem leaves are constrained for reasons of supply, effectiveness or safety.13/

Loss prevention options are thus restricted to a 'high technology' approach in which metal bins are substituted for the bamboo structure.14/ Such bins provide rodent proofing and, carefully manufactured, are sufficiently air-tight to allow fumigation. A detailed Indian case study (Boxall et al 1978) of this option has demonstrated that even at significantly higher levels of food loss the costs of loss prevention outweigh the benefits. Neither the bins nor the fumigants are available in Bangladesh so a comparable study was therefore not possible but there are no plausible reasons for assuming that either input or output prices would be sufficiently more favourable to change the conclusion.15/ The Indian study showed that the use of accounting rather than market prices improved the benefit-cost ratio but at any social discount rate above twelve per cent the loss-reducing innovations still give a benefit-cost ratio below one, even before the costs of extension services have been taken into account. It seems reasonable to conclude that cost-effective loss preventing technical change is not possible in Bangladesh for farm-level paddy storage.

In the pre-storage operations the levels of losses are very much lower; there are no plausible16/ alternative techniques with which to make estimates of benefit-cost ratios. Given the results for raw paddy storage, any such calculation would be academic. This argument is reinforced in a social benefit-cost analysis by the recognition that in the pre-storage operations, especially cutting, gleaners recover many of the "losses".

Gleaners are usually children and women from landless households; people for whom the opportunity cost of labour is very low. Moreover, they only pursue their gleaning with any great enthusiasm in the _aman_ season, especially in the broadest fields where straw is long and there is often standing water resulting in high cutting losses (3 per cent - Table 4.3). These same conditions prohibit mechanisation.

The two operations (threshing and milling) where technical change is actually occurring are discussed in the next two chapters and it is sufficient to note here that in threshing, as has already been shown for milling (Table 4.2), the change of technique actually increases food losses (though in the case of milling these losses are generally recovered - see Section 6.4).

4.7 SEASONS, REGIONS, FARM SIZE AND VARIETIES

These four aspects of a postharvest system are often thought to be critical determinants of the magnitude of physical food losses. In _this_ study, they are relatively unimportant influences on results. This section explains why this is so but it is as well to immediately add the cautionary note, which is elaborated on in the later section on research implications, that these aspects and their interactions are expected to become increasingly important in explaining farmer's attitudes to technical change as land productivity increases.

Central to the discussion of all four aspects is the low standard error of the estimated losses in each operation. These are given in Tables 4.2 and 4.3. A more or less accurate measure of the reliability of results is given by the 95 percent confidence limits of an estimate, which is approximately two times the standard error, (the coefficient of variation is a similar, more familiar but less accurate and interpretable description). In all non-storage operations (except _aman_ season broadcast cutting) these limits represent variation around the estimated mean loss of less than one percent. In storage the range is greater than one percent in half of the eight results (Table 4.3), but only exceeds one and a half per cent in the case of aman storage of parboiled paddy.17/

For any specified experimental design the combination of low losses and low variation indicates: (i) that the influence of 'treatment' variables, such as region, farm size, etc, is small and, (ii) that at any given level of influence, the increase or decrease in value of losses due to a 'treatment' variable is small, reflecting the low average value. 'Smallness' here is of course to be interpreted in a policy sense: ie, given prices and choice of potential techniques the variability in losses associated with any of these four aspects is not sufficient to offer cost-effective intervention opportunities. This is a strong claim and is subject to various qualifications in relation to each of these aspects.

On seasons there are two important qualifications. First, only the aus and boro have been aggregated in calculating standard errors. Inspection of table 4.3 shows that field stacking and storage loss levels do not differ significantly between the aman and aus/boro seasons; in fact, the standard error across all seasons for paddy storage losses is only 0.18. For transport, the seasonal variation is statistically significant but the average difference between aman and aus/boro of 0.33 per cent is unimportant in a policy sense. But, for cutting and for threshing, loss levels do appear to differ by season; these differences however are both also associated with varietal differences in straw length; losses within seasons eg for aman cutting and aus/boro threshing are also significantly different for the same reason. Unfortunately the varietal mixes employed do not allow a cross-season comparison for the same or similar varieties.

Secondly, it has to be emphasised that the results relate only to physical losses and that qualitative losses are more likely, for obvious reasons, to be significant in the wet season (aus/boro) harvest.

Regional differences in losses can be due to use of alternative techniques or due to changed climatic conditions, especially rainfall. In Bangladesh, farm-level postharvest techniques are relatively homogeneous. During the course of fieldwork there was opportunity to study postharvest practices in all but four of the twenty districts in Bangladesh; it was observed that there are small differences in technique in many operations (eg field stacking, transportation, farmyard stacking, storage) but their impact on losses can hardly be significant.18/ In some cases, use of a different technique is probably a response to enhanced risk of losses; in Sylhet, the wettest district of Bangladesh, threshing and farmyard stacking methods are frequently designed19/ to take maximum advantage of dry spells during harvest, in ways which farmers elsewhere apparently do not feel is necessary.

Rice milling, presents the most interesting case for analysis of regional differences in technique because of two features, both however relatively unimportant so far as food loss is concerned. First, traditional rice milling, dominated by the dheki, has two distinct regional variations with the use of the manual pestle and mortar, khol chiya in the North East and use of the dholong 20/ in the South East. Culinary differences may explain these variations though no detailed analysis has been made. Secondly, there are distinct differences in the diffusion of mechanical rice milling; these differences, as argued in Chapter Six, reflect the differential development of the infrastructure and are tending to reduce over time.

Techniques, then, are not a cause of regional differences in loss. But, with given techniques, climatic differences could effect loss levels. In Bangladesh, rainfall variation is the main candidate for analysis. There is a strong rainfall gradient (see Figure Five) with a mean annual precipitation of below 1400 mm in the north west and above 5800 in the hilly north east. But, the gradient is not smooth and two thirds of the country has a mean annual rainfall below 2200 mm (Manalo nd, p.333).

Both the field stations were in regions where the mean annual rainfall was between 2000 and 2200 mm; wetter than most of the country but not nearly as wet as the hill country. In fact, it is difficult to relate field areas in specific years to annual norms because of the high variation. 1979 was a dry year and 1980 was a normal year nationally but in the two field areas rainfall ranged from nearly 3000 mm (1979-80) to 2000 mm (1980-81) (GOPRB 1983a, p.156). Within each year, the most critical factor is the number of days of continuous rain during harvest time; no information is available on this factor. In actual field work however, it was possible, at some cost to random sample design, to work in situations where farmers were facing difficult wet season conditions (see next section).

Overall, the only qualification on the thesis of low regional differences is to acknowledge the variable risk of wet season problems which, as argued below, are anyway not a main cause of reduced food availability.

Neither farm-size nor tenure-status are associated with differences in postharvest technique. The effect of differences in volume of production was largely restricted

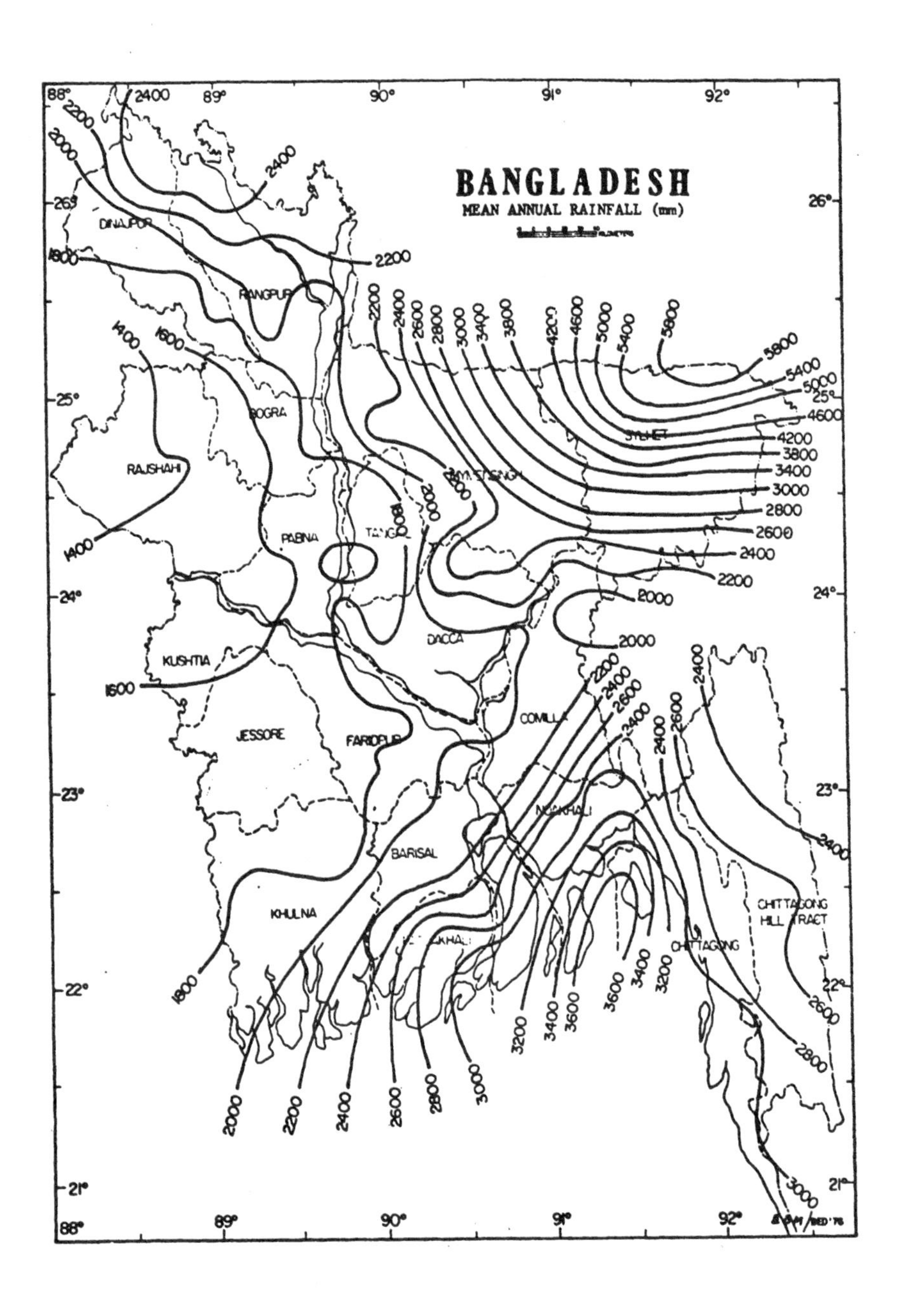

Figure Five (from Manalo, n.d. p 333)

to farmer capacity to cope with the threat of qualitative deterioration in the wet season. This effect was positively related to farm-size because limited farmyard space, bullock availability and farm management time resulted in delays in threshing and extended farmyard stacking - these problems are discussed in the next section on wet season losses. So far as physical losses are concerned, the various arguments21/ suggesting either scale economies or scale diseconomies were not supported by the results.

Average holding sizes in the Comilla and Tangail field areas are reported in Appendix Three along with national averages. Detailed examination of farm size effects was restricted by the exigencies of assessment methods but for loss estimates in cutting to threshing (composite), threshing drying and storage operations, where statistical analysis of farm-size effects was possible, no noteworthy differences emerged.22/ Some, more complex considerations, such as difference in quality of family and hired labour or the likely lower probability of some large farmers completing a second threshing, could not be examined; given the overall low losses, and low CVs, this does little to qualify the assertion of insignificant differences due to farm size. No specific evidence was collected on tenure-status and loss levels but, again, the low losses and low CVs suggests that this will not be an especially fruitful avenue to explore.

On varieties both the cutting and threshing operations require qualification of the earlier generalised assertion that losses and varieties do not interact. Groups of varieties, characterised by differences in straw length, do show variations in loss levels. There are three groups. (A) Traditional deep water varieties, where straw length can be as much as four metres from roots to panicle; (B) non-deep water traditional and locally improved varieties commonly having a cut straw length over one metre; they are both broadcast and transplanted; and (C) dwarf varieties, all of which are HYVs, whose cut straw length is around 0.6 metres. Group C varieties are always transplanted. A series of tests showed that in cutting and threshing, losses are significantly higher both in Groups A and B than Group C; there were no other significant differences.23/

The policy implications of these differences is limited however. The choice of variety group largely reflects the probable depth of flooding and more generally the degree of water control. Group A varieties are grown under rainfed conditions where flooding is expected. The stalk elongates as the water level rises; it elbows,24/ and flowers after the water level drops. By harvest time stalks are commonly thickly intertwined and individual stalks extend from one to four metres, horizontally, depending upon the depth of past flooding. Cutting becomes difficult and, with fields often still wet, higher losses become likely and the accuracy of measurement is reduced. It is not easy to improve water control. In the aman season, where group A varieties are grown, a major change in land use, currently gathering momentum through increased investment in pump irrigation, is the replacement of group A and to some extent group B varieties in the aman with a group C variety grown in the boro season. This change is helping to reduce postharvest losses, though it is also true that group C varieties require more careful farm management; loss levels are more sensitive to timing of harvest operations since late cutting of these varieties can substantially increase losses due to shattering.

Transportation provides an example of a statistically significant but policy irrelevant difference in varietal (and seasonal) circumstances. The group C varieties whilst suffering higher absolute loss per head-load had a significantly lower percentage loss than group B; this is due to the increase in the grain to straw ratio. However, the change was from a loss of one third to a loss of two thirds of one percent - the higher loss level is still too low to merit intervention.

Some HYVs are more susceptible to attack by insects in storage because breeding for high grain content tends to be associated with thinner husks.25/ In this study, only marginally higher losses (3.2 compared to 3.5 per cent) were found when group A and B varieties were compared to group C varieties for raw paddy storage. Sample sizes were too small and numbers of varieties too large to prevent more detailed comparison but the evidence suggests that, at least so far, this varietal characteristic is not significant for food loss.

4.8 WET SEASON LOSSES

The evidence from group one operations is unambiguous. Losses are moderate to very low. Intervention in any operation to prevent those losses is not cost effective. However, as Section 3.5 described, the main thrust of the interventionist lobby has been towards the introduction of farm-level crop driers to prevent qualitative deterioration arising from delayed postharvest operations, especially drying, under wet seasons conditions. The group two operations allow farmers to make an adjustment in practices when continuous rainfall impedes the normal sequence of operations. Extension of the stacking or soaking period, the former being far more common, and delay in drying, alone or in combination, provide flexibility but at the same time create an increased risk of qualitative deterioration in the wet grain.

In response, research teams in Bangladesh during the last decade have tried to promote solar paddy driers, husk-fired paddy driers and various sizes of electrically and fossil-fuel powered paddy driers. None of these have been successful and the evidence from the project wet season study throws some light on why this was the case. Driers offer a clear substitute for only two of the three wet season practices: extended soaking and delayed drying. The principle reason for extended stacking is a rain affected threshing floor delaying threshing operations and crop driers cannot be brought into operation until threshing is complete. Therefore, for farmers using traditional threshing methods crop driers will not provide any solution to the problem of continuous rain preventing threshing; and where weather conditions allow threshing to be performed, they will usually allow traditional sun-drying, obviating the need for crop driers. However, farmers who use pedal threshers can benefit from crop driers - since these threshers can be used indoors; the case for pedal threshers is the subject of the next chapter.

The economic logic of the crop drier is simply based on the hypotheses that (1) inability to pursue adequate and timely drying for paddy gives rise to a loss of economic value to the farmer; and that (2) to avoid this, it will pay sufficient farmers to use a paddy drier for such a drier to earn an adequate financial rate of return. Only two studies have addressed themselves to empirical verification of (1) but they (Refs. 6 and 8 of Table 3.3) have not produced meaningful results. The argument in favour of driers has had to be more loosely constituted along the following lines. It usually rains during the aus harvest and HYV boro harvest. Because of rain, both raw and parboiled paddy drying is delayed. Moist paddy rapidly deteriorates and its value falls; ie, its milling quality and its consumer observable quality characteristics (appearance and smell) worsen. (And it is possible that worrying levels of toxin-producing microflora will result - these are examined in the next section).

To address this argument, the project collected four types of information through farmer questionnaire surveys and sample analysis. First, through a small preliminary round of farmer interviews, evidence on farmer perception of the severity of wet season losses was developed. Secondly, a small set of intensive studies on the relationship between wet season operations and milling yields (and hence losses) was undertaken. Thirdly, a large set of farmers interviews was undertaken to establish their perception of the most problematic operations. This revealed that extended farmyard stacking was the biggest problem, followed by extended soaking and delayed drying of parboiled paddy; in the next wet season a special study of both these operations was completed.

Interpretation And Valuation: Two Qualifications

Before presenting the results of these efforts two qualifications have to be made. First, the distribution of rainfall in the period of harvesting is a critical yet underresearched topic. Rainfall records giving daily precipitation do not reveal the distribution within the twenty-four hour period and different distributions greatly affect the scale of the problem. Continuous light or moderate rain over several days is the main problem yet from the records they cannot be distinguished from sporadic short bursts of heavy rain. Equally indistinguishable, rain at night is less of a problem than rain in the daytime. Ideally, one would have a measure of the probability of rain-induced changes in farmer practices from which to assess the results of the experimental work on the effects of those changes in practices on crop value. No such measure is available.

Secondly, the estimate of change in the value of the crop is a highly uncertain exercise. The treatment adopted here, only valuing physical losses due to reduced milling yield as a result of qualitative deterioration, is consistent with the traditional emphasis regarding qualitative losses. But, a growing trend in the literature is a concern with deterioration in consumer observable quality characteristics independent of quantitative loss.26/ Price sensitivity to rice quality tends to increase as incomes increase. It was noted earlier that quality-sensitivity of prices in Bangladesh is probably small by international standards but even in low income countries such as Bangladesh, some price sensitivity can be readily observed (Nurunnabi and Amanatullah 1980). The exclusion of a valuation for consumer observable quality characteristics therefore requires some further explanation.

The price penalty in the case of such qualitative deterioration as noted in Section 3.2 does not reflect loss in the nutritive value, which is difficult to analyse but anyway very low; rather it reflects loss in attractiveness to the rice eaters. As chapter three argues, deterioration in consumer observable quality characteristics of consumption paddy, though theoretically valued the same as sale paddy, in practice is probably valued at a lower level by a majority of farmers whose net sales if any are a small proportion of their total production. The proposed explanation of this difference was outlined in Section 3.5 and is twofold. First, that paddy identified for consumption may be allowed to deteriorate when the labour and other inputs needed to prevent such deterioration have an opportunity cost greater than the perceived loss consequent to deterioration; this argument is most strong when prevention of deterioration requires cash expenditure (say, on labour for threshing) because, at low levels of cash income, farmer consumer preference for better quality rice will often be less significant than his desire to increase disposable cash income.

Secondly, transactions costs limit substitutability of consumption paddy for sale and purchase paddy. Accepting the first argument that poor farmers have, perforce, a high degree of indifference to consumption rice quality, it would appear that such farmers could sell good paddy and buy back more than an equivalent amount of poorer quality rice. If their paddy deteriorates this is obviously not possible. The realisation of this theoretical benefit from avoidance of deterioration would be through additional crop sales and repurchase, at lower prices, of consumption rice; however the additional transactions costs - retail margins, transport costs and time - reduce this benefit and, very possibly, eliminate it. Transactions costs therefore result in a further weakening of the case for using a market value for qualitative losses of consumption paddy.

A rigorous test of these two arguments is not possible - a parallel problem, which though more familiar is equally difficult to resolve empirically, lies in the relative valuation of family and hired labour in estimating cultivation costs. The opportunity cost of family labour, for any specific agricultural task, is often assumed to be lower than the wage rate for hired labour for a variety of reasons - notably, high search and supervision costs for hired labour.27/ The common theme in these two problems is valuation in a non-market situation. To value qualitative deterioration of a given quantity of consumption rice at the (lower) market price of that rice is akin to valuing family labour at the market wage rate. In neither case is the imputed cash cost actually incurred. (In the next chapter, we return to the family labour valuation problem and argue that there are circumstances where family labour may have a higher opportunity cost than hired labour, but for farm management reasons and not because of its market price).

Evidence that tends to support the recognition of these value differences is provided by a price analysis of the local paddy and rice markets. The impact of quality deterioration on prices was explored in this project by a study that related market price to levels of rancidity. The results, reported in Nurunnabi and Amanatullah (1980b), show that increases in rancidity were significantly correlated with decreases in prices. They provide a basis therefore for estimating the reduction in the imputed market value of deteriorated consumption rice. For the levels of rancidity changes in the fifty pairs of samples analysed, the average price fall was 15.9 per cent for raw paddy and 8.4 per cent for parboiled rice. (This difference in price response was not due to difference in the level of rancidity and, therefore, suggests that there is more consumer observable evidence of deterioration in raw paddy than in parboiled rice.) However, despite this relatively large (8-16 per cent) loss in value, farmer-level driers are not in demand. Why is it that the attempted introduction of driers by researchers has not been successful despite the range of technologies tested, the different institutional approaches and, at this level of losses, economic viability?

We suggest that the two arguments advanced above, concerning differential treatment and differential valuation of consumption paddy, constitute a highly plausible explanation of this behaviour. Cash conscious farmers will avoid cash expenditure for drying unless not to do so would incur higher cash losses - and deterioration of paddy earmarked for consumption does not result in cash losses. Behaviour of this sort was in fact offered as the explanation for the failure of one of the drying projects.28/ It was argued that the reluctance of farmers to bring grain for drying was because of their willingness to tolerate a degree of deterioration; the little paddy that did come to the drier was threatened with complete loss of edibility.

The corollary of this argument is that the potential for crop driers, if any, lies with the marketed surplus.

The implications of this are discussed below in relation to research priorities and methods. The evidence on wet season losses is evaluated here from an on-farm consumption perspective in which only reduced milling yields due to qualitative deterioration in earlier operations are incorporated.

The Results of The Wet Season Study

With these two qualifications, on interpretation and valuation of empirical findings, the conclusions on group two operations are necessarily more conjectural than the 'hard' estimates presented for the group one operations. Nevertheless, they are sufficiently firm to support an argument that whilst the risks of losses are real, the various adjustment strategies farmers follow establish the undesirability of farm-level investment on drying at current yield levels.

The first, very preliminary, study is summarised in Table 4.5. Three end-use changes that involve clear-cut economic loss are associated with marked deterioration in wet season paddy; complete discard, use as feed and sale at a lower price. The infrequency with which these changes occurred over the last five years is evident from the table; only 121 maunds were affected and only 50 maunds suffered complete loss of food value. This complete discard was well below half of one per cent of total production. Such responses were never practised by nearly two-thirds of the sample. This study served the limited purpose of establishing that the extreme measures, discard and use as feed, appear to be rarely practised. In the two field areas (3094 households) over two years no such examples of complete food loss were encountered.

The second survey was a quantitatively based estimate of reduced milling yield associated with different postharvest operations in the wet season. Samples were taken after each operation from cutting through farmyard stacking, threshing, drying (raw), soaking and parboiling and drying (parboiled). 83 samples in all were evaluated using the laboratory rice mill to measure the per cent polished rice and per cent brokens (using fixed milling settings and after each sample had been shade dried to 14 per cent moisture content).

The selection of samples was directed to problem plots identified by farmers. Even though, on average, the 1979 monsoon was mild there were no lack of opportunities to find rain-affected postharvest operations. In five out of 18

Table 4.5

Farmers' Experience Of Three Wet Season Postharvest Loss Situations

A. Frequency of Practice

	Complete discard		Use as feed		Sale at lower price	
	No.	%	No.	%	No.	%
No. of interviews	35	100	35	100	35	100
Never practised	22	66.86	23	65.72	10	28.57
Ever practised	13	37.14	12	34.29	25	71.43
Practised within the last five years	8	22.86	9	25.71	16	45.71

B. Quantity involved in the last five years practice

Variable	Nos Farmers	Quantity (in maunds)	Average quantity (in Maunds)
Complete discard	8	25.12	3.14
Use as feed	9	24.75	2.75
Sale at a lower price	16	71.36	4.46

samples drawn in farmyard stacking, the paddy had been dragged home through the water tied in a long floating chain rather than carried by hired labour on head or shoulder pole load as normal. In eight cases stacking was extended more than four days because of rain. Yet, as Table 4.6 shows, there were no significant differences in the yield of polished rice. A paired t-test, reported in the notes to Table 4.6, comparing post-threshing and post-drying milling yields, provided a direct measure of the effects of drying on milling yields. This comparison, by good fortune, included six cases where continuous rain had forced the farmer to adopt the extreme practice of attempting to dry his paddy on floor space within the home. This is indicative of a severe drying problem, yet the milling tests showed no change in per cent polished rice, though there was a significant increase in the per cent broken grains.

These results do not indicate any one operation very clearly responsible for wet season reductions in milling yield. The next stage was to conduct a questionnaire on perceptions of wet season problems amongst 180 farmers in the two field areas; they cultivated a total of 683 paddy plots in *aus* 1979. The results (Table 4.7) were remarkably clear. Of the four variables covered, i.e. per cent affected: plots, area, yield and farmers, - only one operation consistently had fifty per cent or more cases of a perceived problem. (Only one other operation, cutting, had even a single fifty per cent score - more than fifty per cent of farm households reported a cutting problem - but see the note to Table 4.7). The operation so unambiguously identified by farmers was farmyard stacking. Eighty per cent of farmers and more than fifty per cent of plots, area and yield were affected by delay in stacking. Looking at the causes of delayed stacking, no problem other than continuous rain (Table 4.8) affected more than ten per cent of plots, area or yield.

Only 4 out of 180 farmers referred to drying problems with freshly harvested paddy. This low response reflected the results for drying of raw paddy reported above in suggesting that delayed drying of raw paddy was not a serious source of physical losses. As was expected29/ many more farmers, and one-fifth of all of them, referred to drying of parboiled paddy as particularly problematic.

These results clearly suggested a focus on extended farmyard stacking; the second priority of farmers (Table 4.7) though much less important, was drying after parboiling. There are two components to this problem. Extended soaking prior to parboiling and delayed drying

Table 4.6
The Effect of Wet Season Postharvest Operations on Milling Yield

Operation	No. of cases	Average moisture content	Average % polished rice	Average % brokens
Cutting	13	18.42	66.75	5.70
Stacking	18	19.58	67.44	5.56
Threshing	22	20.30	66.37	7.68
Drying (raw)	13	15.23	68.48	10.75
Parboiling	9	21.64	67.85	2.78
Drying (parboiled)	8	14.74	68.20	9.14
			F=0.37 with 5 and 77 df. insignificant at all levels	F=3.17 with 5 and 77 df. significant at the 95% level

Note: The samples were drawn after each operation as classified; moisture contents reported in the table are from the time of sample collection. The milling results, from the laboratory huller and polisher, were obtained after all the samples had been shade dried down to 14% moisture content. Samples were also drawn after husking and polishing but are not comparable in this test because of the different basis for the milling yield.

The percentage of polished rice is the weight of the rice yield expressed as a percentage of the original weight of paddy. The percentage of brokens is the weight of brokens expressed as a percentage of the weight of the rice.

The significant F-ratio in the per cent brokens is due to the results for parboiling; if we separate these and reconstruct the F-test on the remainder, the obtained F is 1.83 which is non-significant at all levels.

We took 10 lots of grain, where both post-threshing and post-drying milling yields were available, to test for the effect of sun-drying on milling yields in a paired t-test. The results for per cent polished rice were not significant (t=0.33) but there was a significant difference at the 95% level for per cent brokens (t=2.79).

Table 4.7

Farmers' Perception of Wet Season Problem Operations

Operation	Total plots	Affected Plots		Total area	Affected Areas		Total yield	Affected Yield		Total farmers nos	Affected Farmers	
		(nos.)	(%)	(acres)	(acres)	(%)	(Maunds)	(Maunds)	(%)		nos.	%
Cutting	683	325	47.58	193.7	89.94	46.43	2095.24	829.43	39.59	180	106	58.89
Transport	683	12	1.76	193.7	4.22	2.18	2095.24	47.50	2.27	180	9	5.00
Stacking	683	352	51.54	193.7	101.51	52.41	2095.24	1072.55	51.19	180	145	80.56
Threshing	683	29	4.25	193.7	7.93	4.10	2095.24	107.24	5.12	180	26	14.44
Drying (raw)	683	5	0.73	193.7	1.84	0.95	2095.24	30.00	1.43	180	4	2.22
Parboiling	683	9	1.32	193.7	3.62	1.87	2095.24	51.75	2.47	180	7	3.89
Drying (parboiled)	683	47	6.88	193.7	16.96	8.75	2095.24	242.50	11.57	180	36	20.00

Note: Cutting "problems" were defined to include all cases of harvesting when there was standing water in the field, and this accounts for over three hundred cases, where in fact this "problem" is normal. The cutting results are therefore only comparable in that they are the cause of greater labour effort in harvesting, rather than in the sense of a cause of deterioration.

Table 4.8

Causes of the Extended Farmyard Stacking Period Before Threshing

Problems	Total plots	Affected Plots		Total area (acres)	Affected Area		Total yield (Maunds)	Affected Yield		Total Nos.	Affected Farmers		Nos. of plots with pre-mature germi-nation	% of affec-ted plots
	(nos.)	(nos.)	(%)		(acres)	(%)		(Maunds)	(%)		(nos.)	(%)		
Continuous rain	683	131	19.18	193.7	42.85	22.12	2095.24	500.43	23.88	180	87	48.33	43	32.82
Muddy threshing floor	683	56	8.20	193.7	16.94	8.75	2095.24	192.05	9.17	180	34	18.89	21	37.50
Bullock power constraint	683	21	3.07	193.7	4.88	2.52	2095.24	50.25	2.40	180	16	8.89	5	23.81
Labour power constraint	683	4	0.59	193.7	1.80	0.93	2095.24	18.50	0.88	180	3	1.67	2	50.00
Time constraint	683	37	5.42	193.7	9.54	4.93	2095.24	97.12	4.63	180	27	15.00	2	5.40

Note: In addition to these reasons 61 farmers (145 plots) in Comilla reported "achieving uniform ripening" as the reason for an extended stacking period and two farmers reported a space constraint.

afterwards. Before turning to our results on extended farmyard stacking we present our, somewhat limited, evidence relating to this second priority.

A small controlled field test was set up to examine the effects of extended soaking and delayed drying on milling yield. Paddy and rice samples were taken from lots of grain soaked from one to eight days. Two different drying practices were examined; drying immediately after parboiling and drying delayed for two days after parboiling.30/ There was no evidence (Table 4.9) that moderate extension of usual soaking periods would cause deterioration, but extensions over seven days did result in substantially reduced (72 to 64 per cent) milling yields. However, such an extreme measure is very rarely resorted to. The results also suggest that stacking after parboiling can be a problem; however, if the one-fifth of the farmers reporting drying problems after parboiling in *aus* 1979 had all adopted this extreme practice for all their affected yield, just over one-half of one per cent of total yield would have been lost.31/

Extension of the pre-threshing stacking period was examined for effects upon physical loss in two ways; the effect upon threshing losses and upon milling yield. Unlike soaking estimates controlled experiments were neither practically feasible nor likely to faithfully represent farm practice. The approach adopted in the *aus* 1980 harvest was to sample farm stacks to obtain samples for milling tests before using the threshing loss procedures described earlier (Section 4.5) to assess the effects, if any, of extended stacking on completeness of threshing. Twenty sets of observations were made in all from the four project villages in Comilla with the period of stacking varying from less than one day to fifteen days. All the observations were taken from local varieties widely used in Comilla in broadcast mixed *aus*.

It was originally planned to include a parallel set of observations on milling yield for the predominant transplanted HYV, Purbachi. However, stacking was rarely practised for more than a day or so; pedal threshing - universally employed in these Comilla villages for the HYV crop - was not constrained by inclement weather or a muddy threshing floor. This important feature of the pedal thresher is discussed in the next chapter on choice of threshing technique.

In the second field location in Tangail, even though farmers did not have access to pedal threshers, the transplanted HYV *aus* crop was not stacked for long periods before first threshing. However, between first and second threshing extended stacking did occur. (With stiff-strawed paddy the first threshing is performed by hand beating and this is usually followed with a second threshing by animal trampling.) Observations of the effects of this practice on milling yield were difficult to make because extended stacking periods over one week were still relatively uncommon; the four results that were obtained gave a milling yield of second threshed paddy of 64.53 per cent (19.24 per cent brokens) when stacking was more than one week. For the fifteen results where stacking was less than a week the milling yield was 67.12 per cent (9.2 per cent brokens). The value of this apparently substantial loss due to extended stacking is not actually very large once adjustment is made for the fact that only about five per cent of total yield comes from the second threshing and that the extended stacking required to produce such losses was very uncommon. A contributory cause of extended stacking appeared to be poor access to draught-power for second threshing because bullocks were being used for land preparation of the aman crop. This farm-management aspect of the wet season postharvest problem is examined more in the discussion below on research priorities and methods.

A summary of the observations from Comilla on extended stacking of the broadcast crops is given in Table 4.10 which includes average milling yields and average threshing losses for stacks of less than one week and more than one week. The observations were in fact spaced evenly over fifteen days but no greater clarity of observation is achieved by a further breakdown.32/ They show that there was a marginal (non-significant) reduction in threshing losses. This result was consistent with farmers' views that achieving completeness of threshing was easier after extended stacking. The milling results showed no change in the percentage of milled rice but almost double the number of brokens when stacking was extended for more than one week.

Whilst brokens are sometimes used to feed chickens and sometimes in petty barter trade with village hawkers, they are most commonly cooked with the whole rice. As Dawlatana (1980) has shown, through a set of cooking tests conducted as a part of this study (and reported in more detail in Table 6.2) the effect of brokens is to increase the loss of solids in cooked rice because they are dissolved more readily in the cooking water. The food value loss depends on what happens to the excess cooking water. In rural areas this gruel is generally used as food though occasionally it is given to cattle.33/ [Information was not readily

Table 4.9
Effect of Extended Soaking on Milling Yield

Days soaked	No stacking after parboiling		Two days stacking after parboiling	
	% milling yield	% brokens	% milling yield	% brokens
1	72	6.40	68	32.20
2	72	5.27	69	9.87
3	72	6.33	73	10.60
4	72	6.20	70	11.33
5	71	6.90	63	6.88
6	71	7.88	64	5.84
7	63	8.88	63	9.92
8	65	6.96	-	-

Note: After collection (immediately after parboiling or after two days stacking) the samples were dried under controlled conditions down to approximately 14 per cent moisture content.

Table 4.10
Extended Stacking, Threshing Losses and Milling Yield

Stack Period	Per cent threshing loss	Per cent milled rice	Per cent brokens
Less than 1 week n = 10	3.9	68.1	11.9
More than 1 week n = 10	3.4	68.4	20.3

available[34] on how common it is to cook with excess rather than just sufficient water but where excess water is not used there is no reduction in food value.]

To summarise, the results on physical losses due to qualitative deterioration during pre-threshing stacking, soaking and parboiling and drying do not indicate significant losses. A higher percentage of brokens is experienced in only three cases; where stacking, both before threshing (Table 4.10) and after parboiling (Table 4.9) is extended and in the case of sun-drying compared to shade-drying (Table 4.6). However, cooking broken rice does not cause food loss when no excess water is used in cooking. Milling yield reductions were observed where the second threshing was delayed more than one week (p.191) and where stacking of wet paddy after parboiling extended for two days (Table 4.9). These practices are extreme and uncommon adjustments to wet season conditions; they have little claim as a source of postharvest losses that can be cost-effectively prevented.

One apparent contradiction in these results relates to the fact that over eighty per cent of farmers reported pre-threshing stacking problems when the evidence on losses suggests that farmers are not losing substantially as a consequence of these problems. This contradicts the implied view in several earlier sections (eg 2.6) that farmers are sensitive to and perceive accurately the risks of losses associated with particular postharvest operations. The explanation is simple but important; when farmers describe postharvest problems they do so as farm managers rather than as food loss specialists.[35] Thus, when they describe extended stacking as being their most problematic operation it is not just the risk of food loss that accounts for their perception. They also take account of the implications for labour use, draught power use, supervisory time and space constraints when considering the implications of delayed threshing. Extended stacking because of continuous rain that prevents threshing does increase the risks of economic loss through quality deterioration but it is also a bottleneck that upsets planned deployment of scarce farm resources. It is argued in Chapter Five that this farm management factor is an important determinant of pedal thresher adoption decisions.

4.9 THE MYCOTOXIN PROBLEM

Toxin producing fungal flora have been receiving increasing attention in postharvest studies partly in response to a number of well-publicised incidents, mainly concerning groundnuts but some have involved rice, in which high contamination levels caused fatalities (Alpert et al, 1971, Krishnamachari et al, 1975, Peers et al, 1976). The harvest time temperature and humidity conditions in Bangladesh (especially in the wet season) are often close to optimal for development of these microflora and suggest that mycotoxins might well be a serious health hazard. A study was organised to examine this suggestion and is reported in detail by the scientists responsible (Joarder et al 1980). In summary, a two stage screening procedure was adopted. First, nearly 1200 samples were analysed for their fungal flora content. Secondly, those samples where the count of Aspergillus Flavus was over 2000 were analysed for aflatoxin content. 141 samples were analysed. The results are summarised in Table 4.11 (from Joarder et al 1980, p. 52). The incidence rate of aflatoxins was estimated to be 3.9% per cent but only two samples had aflatoxin levels above internationally agreed non-acceptable levels.

The authors, correctly, are very cautious in interpreting these results. Since all the toxigenic strains of A. Flavus are not known, simple use of a population count could have resulted in toxic samples being excluded. Further, there are many other toxins besides aflatoxin which were not taken into consideration. These results are therefore only preliminary and insufficient to make a properly informed analysis of the health risk due to toxin-producing microflora. Nevertheless, they tend to discount alarmist suggestions of the extent of the hazard whilst offering clear advice as to how future studies building on these results should be designed. Of particular interest, the authors demonstrate (Joarder et al 1980, Tables 9 and 10, pp. 38-39) that stacking before threshing and between first and second threshing are causes of increased populations of fungal flora. They support the other results of the wet season analysis in identifying delayed threshing as the chief problem-area in current postharvest practices.

Table 4.11

Incidence of Aflatoxin in Paddy and Rice

Sample		Nos samples examined for fungal incidence (A)	Nos samples[1] containing 2000 above count of A.flavus (per gm) (B)	Nos samples[2] analysed for Aflatoxin (C)	Incidence of Aflatoxin		
					ND	D (D)	NA (D)
Raw paddy (stored samples)	Boro	101	19	13	8	5	0
	Aus	152	56	36	28	7	1
	Aman	634	66	43	35	8	0
Non stored (newly threshed)	Boro	40	22	22	16	6	0
Parboiled paddy	Aus	64	20	15	12	3	0
	Aman	50	2	2	2	0	0
Parboiled rice	Aus	15	6	4	2	1	1
	Aman	128	6	6	5	1	1
Total		1184	197	141	108	33	

Minimum incidence rate of Aflatoxin (%) = $\frac{D}{A}$ x 100 = 2.79

Calculated minimum incidence rate of Aflatoxin (%) = $\frac{B}{C}$ x 2.79

(assuming all the samples of B would have been analysed) = 3.90

1 = counted out of A, 2 = taken from B, ND = not detectable, D = detectable, NA = not acceptable level

Source: Joarder et al (1980) p52.

4.10 CONCLUSIONS AND IMPLICATIONS FOR FUTURE RESEARCH

This study has produced two principal results. First, that in operations where physical loss of food occurs, the levels of loss are too low to permit cost-effective intervention. This result holds true even when different seasons, regions, farm sizes and varieties are taken into account. Secondly, that qualitative deterioration in the wet season is only rarely a cause of milling yield reductions. Several wet season practices result in non-loss deterioration, notably increased brokens and higher mould counts; the most serious problem, identified both by farmer survey and sample analysis, is extended farmyard stacking because of delays in first or second threshing.

The next two chapters on threshing and on milling examine the pattern of existing technical change and, unsurprisingly, show that food loss is not the main reason why farmers innovate. (Labour-saving is). At the low loss levels prevailing, cost reduction can be achieved without any loss reduction. These chapters seek to demonstrate that the emphasis in research on postharvest technical change should be on the employment and income distribution effects of such changes. This leaves open the implications of these results for future loss assessment research in Bangladesh.

If the results related to a static pattern of land use and crop output it would be tempting and probably justified to argue that productivity of future postharvest loss prevention research would be very low and that scarce research resources should be committed elsewhere. However, looking at the results in relation to the patterns of agricultural growth, certain research priorities emerge. Whilst the analysis is inevitably somewhat conjectural, appraisal of the results in relation to both yield increases and changes in cropping patterns support a case for more detailed research on the marketed surplus especially in the wet season. Further, this research should focus on the probable exacerbation of farm management constraints as wet season yields increase. The argument has three steps.

First, an important procedure in the present analysis was to draw a distinction between the valuation of qualitative deterioration of consumption paddy and of sale paddy. Yet, as yields increase, the marketed surplus will increase more than proportionately. In effect, there will be increasing cash income sensitivity to qualitative deterioration. Practices that now are cost effective because they do not lead to cash loss will become less economic. Of special concern, the delays in wet season threshing will become increasingly less attractive temporising measures against inclement weather as the price penalties due to their adoption become more frequently felt.

Secondly, there is already evidence of farmer response along these lines. In Sylhet district, which has a large HYV _boro_ crop harvested under fairly constant threat of rain, several adjustments have already been made, particularly by the large farmers. Threshing is frequently done in the field on bamboo mats not in the farmyard; when pre-threshing stacking is necessary the paddy is stacked in a narrow long line to allow ventilation rather than in the bee-hive shape to prevent theft used in most parts of the country. In addition, more elevated fields close to the farmyard are compacted and used to supplement the farmyard drying area.

Thirdly, the main avenue for increasing output is through the introduction of HYV _boro_ crops grown under irrigated conditions. This crop often replaces a broadcast _aman_ crop, currently the crop suffering the highest level of losses up to threshing. Thus, the change contains a loss-reduction component but at the same time it introduces wet season problems. HYV _boro_ varieties such as _purbachi_ give average yields much larger than traditional _boro_ varieties but harvesting is from May onwards when early monsoon rains threaten. At the same time the farmer is preparing his land for the next crop. Farm assets including bullock power, threshing and drying space and farmer supervisory time all face increasing demand because of the change. Field experience suggests that farmers are faced with difficult trade-offs as a consequence. A good example of these trade-offs is the choice between use of bullock power for ploughing or use of the same animals for threshing so as to reduce the stacking period. Thus, whilst existing farm management of traditional methods works well the capacity to extend these methods as cropping patterns change and yields increase is severely restricted. More tentatively, it can be suggested that large farms (say above five acres) are more likely to suffer as yields increase than other farms - if their assets which are relatively fixed in the short-term, such as farmer supervisory time and threshing floor space, are already being used more productively than on other farms.

A research programme that looked at these concerns would ideally be a component within a farming systems programme since the postharvest problems identified are a result of changes during production. Such an approach would also enable other potentially significant postharvest issues such as thinner-husked HYVs and the wet season toxin problem to share a common focus.

Modest though these conclusions appear, they may be difficult to implement because of institutional constraints. A farming systems approach to postharvest problems has been advocated before (Maxwell 1982) and major farm management programmes have paid at least lip service to this need.36/ However, since postharvest research has so often been conceived of as an exercise in introducing loss-reducing technology the institutional base of postharvest programmes has usually been in divisions of agricultural engineering and not farm management divisions.37/ The man-power and budgetary implications of a shift of postharvest projects between such divisions will certainly cause internal institutional wrangles and very possibly result in less than ideal organisation of future research.

4.11 SUPPORTING EVIDENCE FROM OTHER STUDIES

The results from Bangladesh are quite contrary to the prevailing wisdom concerning farm-level food losses. Recently, however, several other studies have been published which provide further evidence that farm-level food losses have been exaggerated in the past. Some of these results are summarised in Table 4.12.38/ Only the Nigerian study (Giles 1964) is not recent and all but two others were available only after 1980. Tyler and Boxall (1984) survey loss assessment work over the last decade and, regarding storage state (p.9), 'The results from nine of the ten farm loss studies (they reviewed) showed that losses appear to be fairly well contained about or below the 5 per cent level over the storage season.'

Other than the Korean case, the studies reported in Table 4.12 do not measure losses in all postharvest operations, but in no individual operation do losses exceed five per cent.39/ However, they do not always preclude carefully selected low cost innovation, and two of the storage studies quoted (India and Zambia) recommended specific extension efforts to encourage small modifications of existing practices; (none of them suggested priority to the introduction of new types of store such as metal bins). The prohibiting costs in these programmes of modification are in the training and motivation of field staff and village artisans to extend the concepts (of modified cribs, rodent-proof platforms etc.) rather than in the actual cost of construction/manufacture. In practice as with much

Table 4.12

Measured Physical Food Losses in Traditional Postharvest Practices at Farm and Village Level

Country	Commodity	Operation	Per Cent Loss	Note
1. India	Rice	Storage	4.3	Preventable loss 3.2
2. Indonesia	Rice	Standing crop to threshing	3.8	A further 4.5% remains uncut much of which is recovered during gleaning
3. Korea	Rice	Cutting to consumption	11	Preventable loss 7%, storage loss 2-3% (2% preventable)
4. Malaysia	Rice	Standing crop to consumption	7.3	All operations except drying and milling, storage loss 2% - data based on farmer questionnaire
5. Nepal	Rice	Harvesting Threshing	1.2 1.9	Figures reported as incomplete
6. Nigeria	Sorghum and millet	Storage	4	
7. Philippines	Rice	Shattering before and during cutting	below 1	
8. Zambia	Maize	Storage	2-5	

Sources
1. R.A. Boxall et al, (1978, p.58).
2. D. Gaiser, (1981, p.136).
3. C.J. Chung, (1980, pp.4-8).
4. Mohamed, (1981, p.120).
5. S.K. Bhalla and T.B. Basnyat, (1982, p.191).
6. Giles (1964).
7. R.B. Calpatura, (1981, p.373).
8. J.M. Adams and G.W. Harman, (1977, p.4).

'appropriate' technology,40/ the marginality of benefits and the organisational strains imposed on government bureaucracies as well as peasant economy result in little progression beyond pilot programmes. These Indian and Zambian studies both drew attention to the difficulties posed for training and extension by this type of programme and similar difficulties have been identified in Bangladesh, Ghana, Mexico, Senegal and Tanzania.41/

The adoption of metal bins where total capacity is being increased or where large and untypical farmers are the purchasers, often for a variety of non-economic reasons, in parts of India and in some African countries, does not contradict these conclusions. Apart from the subsidies they often receive, there are cost considerations - labour, materials, loading/unloading time, durability, and space/size criteria - which are independent of loss levels. There are also conditions of storage - long periods for soft grains such as wheat in the Indian Punjab - where losses may be on average higher than the results in Table 4.12 and high enough to make moderate investments profitable. (Although the changes in farmer practices required may lead to unanticipated results as happened in Andhra Pradesh, India; farmers using metal bins did not always dry their paddy down to the required 12 per cent as opposed to the usual 14 per cent moisture content and the lack of aeration - unlike in traditional baskets - led to premature germination of seed paddy.) Such opportunities are exceptions however, and the evidence overwhelmingly refutes the stereotyped characterisation of high food losses in farm-level cereal storage.

Of all the postharvest operations, storage has received the most attention in the literature and in R & D programmes, but globally, like Bangladesh, it is threshing and milling where technical change is being most rapidly introduced. Only the Korean study dealt with milling, and the cited milling losses in fact overstate losses attributable to milling per se.42/ Four studies measured threshing losses. These studies, also on rice, reported as follows: Indonesia 2.38 per cent, Malaysia 2.15 per cent, Nepal 1.87 per cent and Korea between 0.5 per cent and 1.65 per cent (our Bangladesh result was 1.79 per cent.) In contrast, the threshing losses questionnaires reported in Table 1.1 were: Malaysia 5-13 per cent, Phillipines 2-6 per cent and Sri Lanka 2-6 per cent.

It is not yet possible to conclude that the naive emphasis of postharvest research on loss reduction has been altered as a consequence of the mounting evidence in favour of more sophisticated analysis. That emphasis will inevitably be tempered also by the growing recognition that the actual pattern of postharvest technical change is not a response to increasing losses but an attempt to reduce costs. This process will be encouraged by the recognition that by posing broader sets of questions postharvest research is not threatened for its survival but actually acquiring renewed analytic importance in understanding the impact of technical change on rural development.

NOTES

1. The study called 'Rice in Bangladesh: Appropriate Technology for the Intra-Village Postharvest System'is described briefly in sections 1.1 and 4.2. Different parts of the results have been reported in many articles and are summarised in Greeley (1981). Unless otherwise specified, all the tables in this chapter are based on data collected during that study.

2. The distinction has become blurred with the introduction of irrigated HYV _boro_ rice which is harvested on average later than the flood-prone lowland _boro_.

3. Stacks do occasionally suffer from combustion; smoking stacks and temperatures beyond the range (60°c) of stored grain temperature measuring rods were common.

4. As yet unpublished data from the FAO-GOPRB Post Harvest Loss Assessment Study currently in progress at the Bangladesh Rice Research Institute.

5. 'I value my garden more for being full of blackbirds than of cherries, and very frankly give them fruit for their songs'. Joseph Addison, in _The Spectator_, No. 477 (c. 1710).

6. Storage loss assessment methods have received considerable analysis notably by the scientists of the Tropical Studies Products Centre as reported in various issues of Tropical Studies Products Information and in Adams and Harman (1977) and also Boxall _et al_ (1978).

7. The 'hill' refers to the group of panicles harvested in one cut with the sickle; the distance between hills and the size of each one depends upon sowing practices.

8. Shattering losses increase as the grain moisture content falls during ripening and this is why farmers try to avoid late harvesting. Farmers drain their fields well before harvesting if it is possible to do so and it is likely therefore that wetter fields will contain wetter grain which is less likely to shatter.

9. Much broken grain results from existing milling practices. The effect of this on the value of losses is discussed below.

10. Storage losses of raw paddy in bamboo baskets in 1978-79 had the following characteristics: mean loss 2.3% (n=36); scores below the mean, 20, scores above the mean, 16; median score, 1.65%; and modal class interval (using class intervals of half of one per cent) 1-1.5 per cent. Storage losses of raw paddy in gunny bags in 1978-79 had: mean loss 3.5% (n=16); scores below the mean 10, scores above the mean 6; median score, 2.25%; and modal class interval, 1.5-2 per cent. For every commodity in every store type in every season the same pattern of skewness existed.

11. Risk of theft is also given as a reason for not storing outside but climate appears to be the main factor. Mphuru (personal communication, 1983, regarding a completed and published survey undertaken by him but no copy yet available to the author) studied African farm level storage patterns and found a very clear association between climatic differences and the materials used and store design.

12. Several deaths have occurred in India and elsewhere due to non-observance of this requirement.

13. A wide range of traditionally-employed natural insecticides or repellants have been studied but despite some positive results their practical efficacy seems limited. Manufactured dusts cannot be mixed with foodstuffs in several countries (e.g. India) because of the health risks.

14. Fully described in Boxall et al (1978).

15. In a social benefit-cost analysis (using border prices) differences in results between India and Bangladesh ought only to occur if there are differences either in the opportunity cost of labour and or in the premium on investment compared to consumption (inter and intra temporal); such differences, if they exist at all, will probably be marginal.

16. Under Bangladesh conditions, options such as mechanical reapers or combine harvesters cannot be considered as plausible alternatives.

17. This was probably due in part to the small sample size (18) and in part because there was greater variability in the average length of storage.

18. Transportation is where there is the biggest variation in techniques, with boats and bullock carts sometimes used instead of shoulder pole or head loads; but this operation anyway suffers very low losses and quite probably the use of boats or bullock carts will tend to reduce these losses by preventing loss due to shattering.

19. Threshing in the field on bamboo mats with the unthreshed paddy in a tight circle round the threshing area and stacking in long rows rather than in beehives are both practices which reduce the risk of deterioration due to sudden rain.

20. All three traditional milling techniques are described in some detail and illustrated in Lockwood (1981).

21. For example, that the larger surface to volume ratio in small stores increases the risk of insect attack in many store types (economies of scale); or that the higher opportunity cost of hired-labour on large farms compared to family labour on small farms results in less labour intensive practices e.g., in threshing (diseconomies of scale).

22. In the case of storage experiments it was possible to use a stratified random sample to identify participating farmers, with the data from the initial village survey. The stratification was three way: those households able to meet their rice consumption requirements from own production (large); those households able to meet at least 50 per cent of their requirements (medium); and those unable to met even 50 per cent of requirements (small). In making these strata, account was taken of production from land leased in or out and of the age and sex composition of the family. In practice, many of the farmers in the "small" category stored for such short periods and used their store so frequently during that period that reliable loss assessment work was difficult and some were excluded; that is why, as noted in the text, the average storage period of our sample farmers was higher than the average for the village.

For other operations, the selection of participating farmers was non-random; decisions were taken the day before, in village meeting places or through visits to farmers' homes and sometimes simply by approaching farmers who were doing the operation assigned to project staff for that particular day. Any other method would have proved impossible to operate since randomly selecting farmers in advance gave no guarantees that the total volume of work could be accommodated in the short time available during the harvesting season given the unpredictability of dates and times for particular operations in particular fields by particular farmers. Nevertheless, because each farmer had a serial number known to the staff, which permitted identification of his village survey proforma, it was possible to stratify ex post; on other occasions, in the second year of fieldwork, selection of farmers was made on a "large" or "small" categorisation on the basis of the local knowledge acquired by the project staff - probably the best basis for stratification even though it was not statistically based.

With the low losses and low CVs that hold for all operations, it was evident just by inspection that no statistically significant differences by farm size were likely to emerge. Nevertheless, for completeness, tests were undertaken on cutting to threshing (composite), threshing, drying and storage; results were non-significant as projected and by way of example one of the tests on threshing is summarised here:

Boro 1980: Effect of farm-size on threshing loss.

t-test for a significant difference between the means

large farms	small farms
$n = 10$	$n = 8$
$\overline{x} = 0.23$	$\overline{y} = 0.37$
$s_x = 0.19$	$s_y = 0.18$

$$t = -1.6$$

$$(df_{16})\ t_{crit} \text{ at the 95 per cent level} = 2.12$$

The t-statistic provides no support for the hypothesis that threshing losses vary systematically by farm-size.

23. The comparison of groups A and C produced similar results to those reported below for groups B and C. No significant differences existed between Groups A and B at the 95 per cent level of confidence, but at the 90 per cent level, the differences in cutting losses were significant.

	Group B (medium straw)	Group C (short straw)
Cutting		
	n = 18	n = 20
	$\bar{x}$ = 2.06	$\bar{y}$ = 0.34
	s_x = 1.50	s_y = 0.37

$t = 4.96$

(df 36) t_{crit} at the 95 per cent level = 2.028

	Group B (medium straw)	Group C (short straw)
Threshing		
	n = 20	n = 38
	$\bar{x}$ = 1.45	$\bar{y}$ = 0.60
	s_x = 1.26	s_y = 0.52

$t = 3.63$

(df 56) t_{crit} at the 95 per cent level = 2.004

24. During flooding, the deepwater varieties elongate with the flood water and float on its surface. 'Elbowing' is the term used to describe the particular genetic characteristic of the deep-water varieties that allows them to grow vertically by angling up after the flood water has receded.

25. See Sriramulu (1973) and Srinavasan (1972). Evidence on the same phenomenon in wheat is given in 'Resistance to Storage Insect in Wheat Grain', Research Bulletin No. 28 of the Indian Agricultural Research Institute, New Delhi, 1980; this study underlines the significance of differential varietal resistance to insect pests but shows that the performance of the newer varieties is not always worse than traditional varieties.

26. See, for example, SEARCA (1983) for an argument along these lines and the review of recent loss assessment studies by Tyler and Boxall (1984). von Oppen (1976) has undertaken a preliminary statistical analysis of Indian consumer preferences for evident quality characteristics of sorghum which provides some useful methodological insights.

27. Another widely accepted argument in support of this view is that much family labour does not regularly seek wage labour work and therefore that the saving of family labour

does not lead to other productive uses of that same labour. Of course, it is equally possible, and in certain circumstances likely, that the non-quantifiable, disutility of family labour effort is rather high.

28. See p.41 in Andersen and Hoberg, 1974.

29. Expected, because parboiled paddy, on average, requires longer to dry.

30. When immediate drying is not possible, farmers stack the hot and wet paddy. The experiment set a fixed two days for this though in practice the period is rarely so long because farmers will not parboil if prolonged rain is likely.

31. 20 per cent of the farmers reported a problem on at least one plot but only 11.6 per cent of the yield was affected. With a 4 per cent reduction (72 to 68%) in milling yield due to delayed drying in the case of paddy soaked for one day, the reduction in output is 5.6 per cent per maund affected. (5.6 x 11.6) - 100 = 0.65 per cent of total yield affected.

32. Excepting that a substantial increase in the percentage of brokens only occurred after stacking for ten days and the results were strongly influenced by one extreme case of 83 per cent brokens.

33. Usually only for the infirm, children, and beggars.

34. In a discussion amongst eight local women from a number of different districts there was no sign of any consensus on what could be called 'normal' practice.

35. Two other explanations that might be suggested are: a) that *male* farmers underassess the problems *female* farmers have during parboiling and drying - but this is improbable for the *male* farmer would certainly be aware of the economic costs of parboiling and drying problems if they were substantial; and, b) that deterioration in stack is highly visible (white moulds, soggy straw etc) and colours farmer perception. However, it is the condition of the grain, not the straw, that chiefly determines economic loss during stacking and farmers are obviously aware that there is no close association.

36. Notably the FAO Farm Management Division (see Greeley, 1983c).

37. In fact, the author knows of no exceptions to this except where Divisions of Entomology also share postharvest interests.

38. The table summarises all those which we have been able to see directly; Tyler and Boxall's (1984) survey of loss assessment includes some other studies which they had access to. Notably, they include several storage results

not listed in Table 4.12. As the text quotation shows, these results tally with those of Table 4.12.

39. The Malaysian evidence is not based upon measured losses but a farmer questionnaire - all the other studies give physically measured losses.

40. See Greeley (1978).

41. Detailed references are given in Greeley (1983b), Footnote 27, p. 534.

42. This is because 'losses' include the reduction in milling yield because consumers prefer highly-polished rice.

Appendix 4.1

Rainfall, Temperature and Relative Humidity (RH) Records in the Two Project Field Areas 1979-80

Month	Madhupur, TANGAIL			Chandina, COMILLA		
	Rainfall mm	Temperature C°	RH %	Rainfall mm	Temperature C°	RH %
June	-	31	75	-	30	78
July	249	29	86	590	30	81
August	338	30	81	380	30	78
September	203	30	83	382	30	79
October	148	29	76	206	30	74
November	8	28	64	400	28	70
December	33	23	67	85	23	67
January	6	20	67	)	22	58
February	2	22	60	)6 months	25	57
March	6	25	61	)total 955	30	53
April	76	32	58	)monthly	33	57
May	395	27	78	)mean 159	30	74
June	337	-	-	)		

Source: Temperature and Relative Humidity: average of three daily daytime readings in project field stations (approx. 9 am, 1 pm and 5 pm). Rainfall: GOPRB (1980b and 1983a) for Comilla and GOPRB (1983b) for Tangail.
NB: the figures given in GOPRB (1983b) are reported as millimetres but self-evidently are in inches; therefore all the reported values have been multiplied by 25.4.
All results rounded to the nearest whole number.

Appendix 4.2

Examples of Field Survey Instructions and Forms

Cutting and Threshing Trials: an outline of methods for hand-beating and cattle-treading threshing.

The notes describe our approach to the estimation of post-production losses from the time of cutting to the completion of winnowing. The methods adopted provide two composite estimates as well as independent estimates of cutting and threshing losses. The composite estimates measure the losses during cutting, stacking, transportation and threshing combined. We shall compare these with the independent estimates of cutting and threshing losses (and at a later stage we shall also be estimating stacking and transportation losses). For each of our composite estimates we shall have two methods of calculation, one based on area and another based on number of hills. In total therefore we shall have six loss estimates. These are described in relation to T_1, T_2 and T_3 and the random sampling chart. These notes describe the field methods. The calculations are similar to those you have made previously except for the hill loss estimates; one copy of each form is attached with a worked example for guidance.

Equipment required:

1. 3 30 meter measuring tapes
2. 1 water-proofed tarpaulin approx. 5 x 5 meters
3. 1 moisture-meter
4. 1 small balance, i.e. 1 kg maximum with gram divisions
5. 1 large balance, i.e. 50 kg. maximum
6. 2 1 sq. meter sampling rods; this has 3 sides of 5-6 mm MS rounds and the fourth side open.
7. 10 pieces of cloth (1 m^2) for carrying unthreshed samples and 10 1 kg sampling bags.
8. Clip board with forms, chart and rough paper
9. Sickles - normally provided by the farmer
10. Marker posts

11. 2 Gunny bags

12. A watch

13. A calculator

1. Select fields that are of visibly uniform yield.

2. Within the farmer's field select an experimental plot, with the farmer's assistance, which is the most uniform; this experimental plot is to be a 9 x 12 meter rectangle which can be measured precisely using the three tape measures as in our earlier work. Mark the boundaries with the marker posts.

3. Always allow cutting to proceed into the field before selecting the experimental plot to avoid edge effects.

4. Measure carefully an area of 10 square meters in the centre of the experimental plot using two tape measures and a sampling rod on either end of the strip. Cut this area first, glean and count the numbers of hills. Thresh on the tarpaulin, clean the grain, weigh it and take the moisture content; record the results on Form T-1. Take a kilogram sample labelled 'central' with the experiment number and date. We have introduced this step because some evidence suggests small samples overstate yields. By following both methods, 10 x 1 sq. meter and now also 1 x 10 sq. meter, we can verify this. This farmer's experimental plot will now consist of the two plots on either side of our ten square meter central plot.

5. Ask the farmer his estimate for cutting time for this experimental plot.

6. Allow cutting to proceed as and where the farmer and labourers wish and, based upon the estimate of time taken, select ten random square meter sample plots according to the chart.

7. Use one sampling rod to measure the square meter sample plot. Be careful to place the back edge of the rod exactly midway between the last cut and the first uncut plant with the middle of the back edge aligned to the

last uncut plant. Use the other rod held vertically to complete the square and cut the plants within the area. Either side edge may go through a plant in which case only include that part of the plant on the inside of the rod. This use of the rod is probably the most difficult part of the experiment and great care must be exercised particularly with direct seeded crops.

8. Inspect the one square meter plot before removing the rod, count the number of hills and collect any broken heads or any grains lying within it. Add these to the sample cut. Wrap the samples in cloth pieces for taking back to the field station.

9. When back at the field station we shall hand strip each sample plot yield on the tarpaulin, carefully inspecting the straw to ensure there are no grains remaining, weigh the grain and take the moisture content. The stripped straw is then collected in one bundle and dried in exactly the same manner as the farmer's unthreshed straw. (See point 15). Record these results on Form T-2 and complete the calculations. Take a composite sample of one kilogram labelled 'sample' with the experiment number and date.

10. Carefully count the number of hills in the whole experimental plot.

11. When cutting is complete use the two rods to mark off an area of $2m^2$ and collect all the grains and broken heads that you find there. This area should obviously not be our central plot or any of our sample plots which have already been gleaned. Repeat this five times. Collect the heads/grains in a gunny bag and take them back to the field station. Thresh, winnow and weigh these grains, adjust the weight to its equivalent at 14% MC for 1 square meter using Form T-1. We are not anticipating great accuracy in this result but over a number of experiments it should provide some indication of the cutting loss element in the total composite results.

12. Let the farmer proceed with transportation, pre-threshing drying, threshing and winnowing in his normal manner. This may involve threshing two or three days after cutting. You must ensure that the straw and grain from the experimental plot is kept completely

separate from that of the remaining field area not included in the experiment. Weigh the threshed grain and take the moisture content recording the weights separately on form T-1 where a second threshing is undertaken. Record details concerning threshing practice and straw length on form T-3. Take a one kilogram sample labelled 'Farmer' with the experiment number and date.

13. Point 11 has provided an estimate of cutting loss and the remainder of points 1-13 have provided a composite estimate of cutting, stacking, transportation and threshing loss. An independent estimate of threshing loss is provided by weighing a randomly selected proportion of the farmers straw (giving F on form T-3). This is taken back to the field station in a gunny bag before it is stripped and winnowed leaving a weight of sound grains (Z). This is adjusted to the equivalent weight at 14% MC, giving L; the formula here is $L = Z \times \frac{(100\text{-Actual MC})}{86}$.

14. In order to convert L into an area specific loss we must know from what area the weight of farmer's straw came. We know that the combined weight of our straw came from 10 square meters, so by weighing our straw (X) we can calculate the loss per square meter implied by L, using the formula on form T-3. Then complete the remaining calculations. It is important that our straw and the farmers straw have been dried in exactly the same manner or differences in straw moisture content will make the loss estimate inaccurate.

15. It should be remembered that these notes are only an outline and do not detail all the small complications that exist in what is a rather difficult experimental design. Only by thinking carefully and working hard at maintaining a co-operative spirit with the farmer will the experiment be successful. Over a number of experiments, about twenty in each area, the method outlined above will provide accurate and reliable loss estimates if all the small details, which are so important in scientific experimentation, are observed carefully and all measurements are made and recorded accurately.

Cutting, Transport and Threshing Trials: Random Sampling Chart

Use with form T-2

Sample number	Estimate of cutting time: minutes				
	60	90	120	150	180
1	5	5	6	7	15
2	9	15	18	23	28
3	13	23	26	34	35
4	23	30	37	54	61
5	29	43	50	65	66
6	31	47	56	80	86
7	41	58	73	93	105
8	46	69	80	107	114
9	53	75	93	125	143
10	55	84	100	136	153

Notes: 1. Start timing from the start of cutting and take your 1 sq. meter samples from the appropriate column at the minutes after start of cutting as given.
2. An allowance has been made for the tendency for cutting time required to be overestimated. Nevertheless you may find that you have to shift to a shorter interval between samples if the estimate was very inaccurate.
3. An allowance of one minute has been made for delay caused by taking our sample cut; if more time is taken you must allow for that.

IDS Postharvest Project: Cutting and Threshing Losses
Farmer's Plot and Central Plot Yields

Exp. No. Example

Form: T-1

F.O. Rahman

Aman 1979

Date: 1st November

Village: Idlepur | Farmer: R.I. Bhuyan | Variety: Pajam | Date sown: 10th August

Field condition: Dry | Crop condition: Upright | Nos of cutters: 2 | Estimated cutting time: 3 hours

Transportation method: Headload | Distance to Threshing Yard: 300 metres | Time between cutting and threshing: 20 hours | Overnight: Yes/No

Moisture contents					
	Prot.	Temp.	Act.	Av. MC	Plot size = 108 M^2 Central cut = 1 x 10 M^2 Sample cut = 10 x 1 M^2 Farmers area = 88 M^2 Gleaned area = 5 x 2 M^2
Before cutting	17.1 16.8 17.4	35 35 35	14.54 14.28 14.79	14.54	
Central 10 M^2	20.2 20.3 20.4	34 34 35	17.37 17.46 17.34	17.39	(Note: Glean this area carefully and include in yield) Yield (grams) = $7154 \times \frac{100\ MC}{56}$ = 6872 = Yield at 14% MC
Gleaned grains from 5 x M^2 (random)	20.9 22.0 21.5	33 35 34.5	18.18 18.70 18.36	18.42	Yield (grams) = $\frac{114}{10} \times \frac{100\ MC}{86}$ = 10.81 = Yield at 14% MC
First threshing	13.3 19.2 19.1	32.5 32 32.5	16.88 16.89 16.71	16.83	Yield (kgs) = $56 \times \frac{100\ MC}{86}$ = 54.16 = A
Second threshing	26.0 23.9 26.9	33 31 30	22.62 21.27 23.73	22.54	Yield (kgs) = $2.8 \times \frac{100\ MC}{86}$ = 2.52 = B

Farmers yield = A + B = $\frac{54.16 + 2.52}{88}$ x 1000 = 644.1 (= C) = Yield in grams per M^2 at 14% MC

Total number of hills = 4300	Number of hills Central 10 M^2 = 409	Yield per hill Central 10 M^2 = 16.80 (CYPH)	$\frac{\text{Central Yield}}{10}$ = 687.2 = Yield per M^2 at 14% MC

Research undertaken by the Institute of Development Studies, Brighton, UK, in collaboration with BCSIR and BARC, Dhaka.

IDS Postharvest Project: Cutting and Threshing Losses

Yield of Sample Plots

Exp. No: Example

PO: Martin

Form T-2

Date: 1st November

Aman 1979 Village: Idlepur Farmer: R.I. Bhuyan Variety: Pajam

Sample	Net weight (grams)	Moisture Content			Act. Av.	Wt. 14% MC Sample yield	Nos of hills	7 8 Hill yield
		Prot.	Temp.	Actual	Act. Av.			
1	2	3	4	5	6	7	8	9
1	769.2	19.0 19.0	28.5 28.5	17.39 17.39	17.39	738.88	45	16.42
2	701.2	19.5 19.1	30.5 30.5	17.45 17.09	17.27	674.54	38	17.75
3	638.1	18.7 18.9	30.5 30.5	16.74 16.92	16.83	617.10	34	18.15
4	692.2	18.9 18.8	28.5 28.5	17.29 17.20	17.25	666.04	40	16.65
5	679.0	18.8 19.0	30.5 30.5	16.83 17.01	16.92	655.95	39	16.82
6	739.1	19.5 20.2	30.5 30.5	17.45 18.08	17.77	706.70	44	16.06
7	637.5	19.0 19.1	31.5 31.0	16.91 17.00	16.96	615.56	37	16.64
8	688.5	18.9 19.6	30.5 30.5	16.92 17.54	17.23	662.64	42	15.78
9	688.7	19.4 19.3	28.5 28.5	17.75 17.66	17.71	658.99	38	17.34
10	734.2	18.8 18.9	30.5 30.5	16.83 16.92	11.88	709.61	44	16.13
Total	6967.7	Notes. Use this space to record extra details that you think may be important.				6706.01	401	167.74
Average	696.77					670.60	40.1	16.77

Research undertaken by the Institute of Development Studies, Brighton, UK, in collaboration with BCSIR and BARC, Dhaka.

IDS Postharvest Project: Cutting and Threshing Losses
Threshing Loss and Summary sheet

Form: T-3 Exp. No. Example

Aman 1979 F.O. Martin

Village: Idlepur Farmer: R.I. Bhyvan Variety: Pajam Date: 1st November

Weight of sample = 4400 (=X) straw dry (grams)	Weight of farmer straw dry: = 3400 (=F) (random sample) (grams)	Straw Length = 72 (cms)
Threshing floor condition: Dry	Details of threshing method(s) and labour/bullocks used:	1st hand - 2 labourers x 1 hour 2nd bullock - 1 labourer x 1 hour

Weight of sound grain from farmers = 6.6 (=Z) Straw: (grams)	Prot.	Temp.	Act.	Av.	Weight at 14% MC
	23	33	20.01		
	23	33	20.01	19.96	6.14 (=L)
	22.7	32.5	19.86		

Area Loss Estimates

1. $\dfrac{L}{10\ \dfrac{(F - Z)}{X}} = 0.80 \times \dfrac{100}{\text{sample yield } (= 670.60)} = 0.12 = \%$ threshing loss

2. $\dfrac{\text{Sample Yield - Farmers Yield (C)} \times 100}{\text{Sample Yield}} = \dfrac{670.6 - 644.1 \times 100}{670.06} = 3.95 = \%$ total loss

3. $\dfrac{\text{Central Yield - Farmers Yield (C)} \times 100}{\text{Central Yield}} = \dfrac{687.2 - 644.1 \times 100}{687.2} = 6.27 = \%$ total loss

4. $\dfrac{\text{Gleaning Yield} \times 100}{\text{Sample Yield}} = \dfrac{10.81}{670.6} \times 100 = 1.61 = \%$ gleaning loss

Hill Loss Estimates

Farmers Hills = Total Hills (4300) (T-1.) - Central Hills (409) (T-1) - Sample Hills (401) (T-2) = 3490

$\dfrac{\text{Farmers Yield (A + B)}}{\text{Farmer'S Hills}} \times 1000 =$ Farmers Yield per Hill in grams (FYPH) = 16.24

5. $\dfrac{\text{SYPH - FYPH}}{\text{SYPH}} \times 100 = \dfrac{16.77 - 16.24}{16.77} \times 100 = 3.16 = \%$ total loss

6. $\dfrac{\text{CYPH - FYPH}}{\text{CYPH}} \times 100 = \dfrac{16.80 - 16.24}{16.80} \times 100 = 3.33 = \%$ total loss

Research undertaken by the Institute of Development Studies, Brighton, UK, in collaboration with BCSIR and BARC, Dhaka.

Appendix 4.3

Percentage Size Distribution of Land Owned (Acres)

<table>
<tr><th>Location</th><th>No land</th><th>0.1-0.49</th><th>0.5-0.99</th><th>1.0-1.49</th><th>1.5-2.49</th><th>2.5-4.95</th><th>5-11.99</th><th>12+</th></tr>
<tr><td>Comilla (1638 households)</td><td>21.8</td><td>30.8</td><td>21.9</td><td>10.1</td><td>9.3</td><td>4.6</td><td>1.0</td><td>0.1</td></tr>
<tr><td>Tangail (1456 households)</td><td>38.9</td><td>15.1</td><td>14.4</td><td>9.4</td><td>11.4</td><td>8.3</td><td>2.0</td><td>0.1</td></tr>
<tr><td>Bangladesh (12×10^6 households)</td><td>28.8</td><td>19.0</td><td>14.2</td><td colspan="3">30.2</td><td>6.1</td><td>1.7</td></tr>
</table>

Source: Comilla and Tangail, the preliminary IDS postharvest project survey in eight villages; Bangladesh, the Land Occupancy Survey, Jannuzi and Peach (1980).

It is worth noting that the land occupancy survey, which is widely quoted, has significant differences to the agricultural census (GOPRB 1980c) conducted in the previous year. For example the land occupancy survey estimated that there were 3.9 million rural households owning some but below one acre of non-household land whereas the census (so called, it was actually a sample survey) estimated there were only 0.99 million households owning or leasing in some but below one acre of farm land. Moreover, this incredible difference is not explained by differing identification of the sampling frame since the percentage distributions are also very different. As the table above shows, 33.2 per cent of households own some but below an acre of land; it can be simply calculated that amongst <u>landowners</u> 46.6 per cent own less than an acre; the comparable figure, including tenants, from the census is 15.9 per cent. These differences remain unexplained. However, we have elected to use the land occupancy survey figures even though their sample size (34,852) was only 8 per cent of the census sample size because they are more consistent with total population statistics than the census results. However, until differences, such as these reported here, are resolved both sets of figures must be treated with caution since the land occupancy survey also had some strange results as is evident from the summary report prepared by Rabbani, Jannuzi and Peach (GOPRB 1978a); for example in Table XIII, 6.8 million households are reported to be growing an <u>aman</u> paddy crop but according to Table XIV only 4.9 million households grew paddy in any season of the year.

5

Technical Change I: Pedal Threshing

5.1 INTRODUCTION

This is the first of two chapters that analyse the economic consequences of technical change now occurring in the Bangladesh farm-level postharvest system. These changes involve the substitution of traditional threshing methods by pedal threshers and the replacement of the *dheki* by rice mills (Chapter Six). The evidence of Chapter Four established that in Bangladesh food losses in these operations are low; loss prevention does not offer a cost effective motivation for adoption of new methods. With no additional output, the benefits of new postharvest technology cannot be evaluated in the manner of other agricultural inputs (water, fertiliser, farm power) by making an assessment of output augmenting effects relative to additional costs. The benefits lie only in the cost reducing characteristics of the technical change. The central problem for economic analysis is to assess how these benefits are shared between labourers, farmers and owners of the new technology.

Two basic features are common to the introduction of pedal threshing and rice mills. They both result in labour displacement and they both operate through custom-hire of the new technology. Labour displacement is an inevitable result, given low food losses and the high labour intensity of traditional postharvest operations (Table 3.5). Custom-hire operation is not inevitable but is unsurprising given the small size of most Bangladesh farms and the indivisibility of the pedal thresher and rice mill. These two features are the principal determinants of the distribution of costs and benefits associated with the new techniques.

5.2 PEDAL THRESHERS: MANUFACTURE, DISTRIBUTION AND USE

The Bangladesh Pedal Thresher consists of a small (16" to 24") rotating drum which has metal loops on its outer surface. The drum, hung on a wood and metal frame, is rotated by one or more operators working a foot pedal attached to the drum by pinions. A manufacturer's detailed cost estimates are given in Appendix 5.1. The direct labour content is only 18 per cent. The market cost (1980) is Tk 1025.

The method of threshing paddy using combs can be traced back nearly three hundred years (Grist 1975, p. 165). Pedal threshing technology has been available in Bangladesh for over twenty years but it is only in the last five years that substantial diffusion has occurred. The Comilla Co-operative Karkhana, an agricultural implements manufacturer, established in 1950 by Dr. Akhter Hameed Khan, has been the single most important influence in promotion of the pedal thresher. There were only four manufacturers in the country in the mid-70s; in 1980, Chandina alone had more than that. The Comilla Co-operative is the largest manufacturer, with sales of over 1500 units in 1979, and has developed a plan to sell over 60,000 units in the current five year plan.

The main reason for the recent acceleration of diffusion appears to be the spread of short stiff-strawed rice varieties. The long thin-strawed traditional and locally improved varieties get tangled in the loops of the drum and their uneven straw length gives poor threshing efficiency. In the Chandina, Comilla field area (1980), all traditional aus was threshed by bullocks and all HYV aus was pedal threshed. The regional pattern of diffusion is therefore associated with the distribution of dwarf and semi-dwarf HYVs. Influenced by the Co-operative, Comilla has the highest density of users but Chittagong, Noakhali and Sylhet, for the boro crop, and the North West districts mainly for HYV aman, are very high growth areas.

Adoption has been aided by three technical characteristics. First, the pedal thresher weighs only 50 kgs and is easily transported; the usual method is to have the thresher slung from its carrying loops on a bamboo pole carried on the shoulder by two men. It is also light enough to be carried on small country boats and by bicycle rickshaw. For longer distances, for example, when taken to other districts by migrant harvest time labourers, it is carried on the roof of a bus. This mobility has allowed diffusion through custom-hire operation on farmers' own threshing yards, where winnowing can still be done by family farm women. Whilst some large farmers purchase pedal threshers exclusively for their own use, the usual pattern of use is for the machine owner or his son and another labourer to move with the machine from farm to farm. Payment for hiring the men and machine is in paddy at the rate of 1 seer for every maund threshed which is a rate of 2.5 per cent; (clearly, there will be some regional variations in contract arrangements but this type of contract was universally employed in the project survey areas). The usual practice involves one or two male family labourers of the farm hiring-in feeding unthreshed paddy to the labour force provided by the machine-owners who operate the pedal thresher.

Secondly, no changes in cutting, farmyard stacking or winnowing methods are required. Thirdly, the simplicity of skills needed for manufacture has allowed very small workshops to start producing pedal threshers, often on a seasonal (harvest-time) basis. The only difficult quality control problem these workshops face is in casting the pinions; however, the Comilla co-operative and another Comilla workshop have surplus capacity on their milling equipment, which cuts rather than casts the pinions, making them more durable. Through the spares market these parts are available for other manufacturers and farmers.

5.3 PEDAL THRESHERS CAUSE HIGHER FOOD LOSSES

If higher levels of food losses were in fact a substantial cause of concern to farmers it would be reasonable to expect technical change to reflect this by preventing some of those losses. The Bangladesh project conducted a set of loss assessment studies to examine precisely the effect of pedal threshers on food losses. The results are summarised in Table 5.1.

The table distinguishes between broadcast and transplanted paddy and, for transplanted paddy, between long and short strawed varieties. This is necessary because the pedal thresher is not used on broadcast crops1/ and because, like traditional methods, it operates at a lower level of threshing efficiency on long-strawed varieties. The table shows that, using the pedal thresher, losses increase by a factor of two on long-strawed and by a factor of three on short-strawed varieties compared to hand-beating follwed by bullock-treading.2/ Use of pedal threshers is much more common with the short-strawed varieties for which losses are still below two per cent.

Table 5.1

Effect of Threshing Technique on Food Losses

Method	Per cent loss	Sample size	Standard error
Bullock treading (broadcast)	2.54	40	0.29
Hand beating followed by bullock trading (transplanted)			
- Short straw	0.60	38	0.08
- Long straw	1.45	20	0.28
Pedal threshing			
- short straw	1.82	10	0.39
- long straw	2.97	9	0.27

Note: Average straw lengths were 1.08 cm (long) and 68 cm (short)

Source: IDS project survey in Chandina, Comilla.

It is clear from these results that food loss prevention is not the reason for farmer adoption. However, the increases in losses, 1.2 per cent for short-strawed and 1.5 per cent for long-strawed varieties are modest; certainly, they are not sufficient to inhibit farmer adoption of the new technique. Moreover, it is possible that with minor modifications (use of a hood or similar device to stop scattering) and greater operator skill as experience is gained (the machines involved in the loss-assessment work had been in use for one to four seasons) these additional losses will be reduced or eliminated.3/

Differences in product quality, if they exist, can have great influence on the most profitable choice of technique (Stewart 1978, chapter nine). In the case of threshing rice in Bangladesh the quality of the milled rice (percentage whole grains) is the relevant parameter. To measure the effects of threshing technique requires carefully organised measurement because rice quality is, of course, influenced by other factors, notably choice of rice milling technique. However, in a set of samples drawn immediately after threshing and which were then treated identically through to milling in the project laboratory, no differences in the quality of milled rice were observed due to threshing technique.

5.4 RETURNS TO PEDAL THRESHER OWNERSHIP

There are three possible benefits for investors in pedal threshers: a) costs reduction in own crop threshing; b) machine rental from hiring out to other farmers; and c) labour-earnings opportunities through provision of labour to operate machines rented out. Of these benefits the most significant is undoubtedly machine rental.

a) Out of twenty machines surveyed in detail only one was bought by a farmer principally for his own use. On average (see Table 5.2 below) there is a labour saving of eighteen hours per ton using the pedal thresher. Assuming an eight year life, the annualised capital cost of the pedal thresher including repairs is Tk 265 at a discount rate of twenty per cent; it can be calculated that a farmer with one crop a year, at the average HYV yield level in our project villages of 50 Maunds (nearly 2 tons) per acre, would require less than four acres to make the investment a paying proposition.4/ With the lower-yielding, longer-strawed varieties, with a higher discount rate or with a lower imputed wage rate the acreage requirement will increase but these calculations are academic for the principal source of income is from hiring-out the machine. It is only the very largest farmers who make the pedal thresher investment purely to cut down own farm costs. However, in considering the benefits from hiring out, discussed next, it should be borne in mind that they will be supplemented by whatever cost savings are made on the owner's own farm.

b) In the project villages there was a uniform pattern of kind payments for pedal threshers at the rate of 1 seer of paddy for every maund of throughput. This rate is two and a half per cent and was always paid by the farmer hiring in to the machine owner at the time of threshing. However, the proportion of this payment that constitutes pure machine rental payment was variable because of the differing labour arrangements used. Clearly, where the farmer hiring-in the machine provides all the labour, including machine transport, the whole payment constitutes rent but this case was extremely uncommon. The regular practice was for one or two machine operators to be provided by the owner of the machine. The payment for these operators was included in the two and a half per cent payment. Where two labourers were provided the payment was divided into three equal shares - one-third to each labourer and one-third for machine rental. The implications of various labour arrangements are discussed in more detail below, but for the purposes of estimating the return on capital, we use the value of a one-third share of the kind payment as the pure machine rent.

Thus, as machine rental, for threshing a ton of paddy the machine owner received 8.3 kilograms out of a total payment of twenty-five kilograms. At a machine purchase cost including transport of Tk 1050 (1980)5/ and the minimum *aman* 1980 harvest paddy price of Tk 3 per kilo, threshing of 43 tons of paddy would be sufficient to recover the initital investment costs.

The quantity threshed each day and the operating days per season are extremely varied. Out of the twenty machines surveyed in detail during the *aman* 1979-80 and *boro/aus* 1980 harvests the investment recovery period varied between three weeks and two years (four harvesting seasons). At the manufacturer's estimated daily (8 hour) throughput of 1.8 tons, which in fact was sometimes exceeded (due to much longer hours which compensated for the lower than estimated output per hour), twenty-four days would be sufficient for full recovery; this suggests that one harvest season may be a reasonable estimate of the average payback period. Maintenance costs do not alter this result significantly. Out of the 20 machines surveyed, eleven had required maintenance or repair expenditure that season.6/ The costs varied from Tk 20 to Tk 150 and averaged Tk 60, or 5.7 per cent of purchase cost.

These figures show that pedal threshing investment is very profitable. A more formal analysis of profitability would require estimates of machine life and of the owners' opportunity cost of capital. For example, with rental earnings of Tk 1,000 per season, two seasons per year, a machine life of ten years and capital at 20 per cent the benefit cost ratio is over 8:1. Such precision is, of course, spurious. Simple machines such as the pedal thresher are repaired endlessly and last indefinitely. There is an active spares and repairs market and a second-hand market. Also, whilst it would be inappropriate here to review the methods of estimating the private opportunity cost of capital,7/ the discount rate used, 20 per cent, is almost certainly an underestimate for rural Bangladesh. Alam (1981) found the rate of interest, including transactions costs, for loans between Tk 1000 and Tk 1500 was between 17.4 and 23.6 per cent for four institutional sources of credit. Most rural borrowing in Bangladesh is from non-institutional sources for which interest rates are known to be much higher.

c) The returns to the investing household (rather than to its capital only as in the discussion so far) can be considerably higher depending on the pattern of labour use. Ownership of the pedal thresher permits the owner a degree of control over whose labour is used, and paid for operating the thresher. The typical pattern of labour use is for one or, much more commonly, two labourers to operate the machine and for one to three labourers to feed unthreshed paddy bundles to these operators. It would appear that the machine owner can choose who will actually operate the machine and that the farmer who is hiring in chooses the labourers who feed paddy bundles. Out of twenty machines surveyed only two were operated by labourers selected by the farmer hiring in and in both cases there were special reasons.8/ The machine owner, by controlling the machine operating labour, can help ensure that the machine is not misused thereby limiting his repair bills and, by providing experienced labour, he can improve the quality of threshing; crucially, he can also control machine utilisation directly and limit stoppages which will reduce his earnings.

In 15 out of the 18 regular cases the machine owner himself, his son or his brother operated the pedal thresher. In 8 of these 15 cases a second member of his family also helped to operate the machine; in four cases the machine owner or his son operated the machine alone and, in three cases, the machine owner also brought with him one of his hired labourers. In the other three cases the machine owner hired his machine out to his own permanent farm servants or to labourers who then undertook contract work for other farmers. In these cases the owner simply obtained the machine rental at the rate of one-third of the total 2.5 per cent kind payment.

The share of the owning households in the kind payment was therefore:

All	12 cases
Two-Thirds	3 cases
One-Third	3 cases

The shares not received by the machine owning household were earned by hired labour selected by the machine owner. The 12 cases where machine owners received all the payment include the 4 where the machine owner worked alone; this is less efficient, since, with sufficient feeders, two operators thresh twice as much as one. The returns to labour are about the same but the returns to the machine per unit time are halved.

Obviously, there is an opportunity cost to the family labour the machine-owner provides and the net returns to his household are less than the money value of the hired payment. How much less is unknown, it will depend on several factors including family structure, labour requirements on the owners' farm and other earnings opportunities. The opportunity cost of family labour is discussed more in the next section but it should be noted here that the money value of kind payments for machine operating labour is greater than the rural agricultural wage rate (see Table 5.2 below); this must be due to the degree of monopolistic control the machine-owners have over operating labour supply and whatever minor skills are associated with operating the pedal thresher. Overall, it is clear that there are substantial benefits from machine ownership.

Given this conclusion it is strange that pedal threshers have not been adopted faster and more widely in the last two decades. One can only speculate on the reasons for this but three causes seem likely. First, that there was a chicken and egg problem. Potential suppliers didn't manufacture pedal threshers because of a lack of demand and potential users were unable to evaluate the technology because they did not know enough about it. The Comilla workshop that first made these machines is located in a town-centre cooperative of urban-based metal workers and carpenters which may have heightened this problem; the workshop itself was frequently chronically short of capital and devoted little time or money to marketing.

Secondly, whilst the acreage under HYVs has grown from under one million acres in 1970 to over five million acres in 1981, only 21 per cent of gross cropped rice area was under HYVs in the latter year (IBRD 1983, p.22). The HYVs are the short stiff-strawed varieties for which the pedal threshers are ideally suited and demand for pedal threshers is closely linked to HYV growers. Demand for pedal threshers is now mushrooming, as are suppliers and it seems that, perhaps for reasons related to the chicken and egg problem, this technology is a later addition to the package of practices associated with use of the HYVs.

Thirdly, and related, the spread of HYVs is the only substantial source of employment generation in Bangladesh agriculture and also results in changed seasonality of labour use. Specifically, the spread of irrigated HYVs during the boro season has led to a major increase in labour requirements during the boro harvest. Since the HYVs are sown later than local boro they are harvested at the same time as early aus and at the same time as sowing of transplanted aman occurs (see Figure Four). In many households this period is replacing the aman harvesting season (December) as the period of peak labour demand. In 1972 the December wage rate exceeded the May rate in all districts except one, Bogra, where they were the same. In 1979 in only one of the twenty districts was the December wage rate greater than the May wage rate (GOPRB 1980b, pp.582 and 589). Therefore, even in a climate of general real wage decline, the relative cost of boro/aus HYV harvest labour has been rising; this trend must certainly have influenced farmer perception and encouraged use of the pedal thresher.

Whether it is these or other reasons that explain the absence of widespread pedal thresher use in the 1970s there is no doubt that in the 1980s the benefits of owning one are clearly and widely perceived. The North-West districts, with higher than average numbers of large farmers and Chittagong, Noakhali and Sylhet, with large areas under HYV boro are rapidly growing markets for pedal threshers and in some cases are beginning to rival Comilla as manufacturing centres.

5.5 RETURNS TO FARMS HIRING PEDAL THRESHERS

The profitability to the farmer who contracts the machine and operators depends upon the difference in costs between traditional threshing methods and pedal threshing. These costs consist of two elements: paid out costs and own farm costs. Survey results showed that, in traditional threshing, the size and composition of these costs are sensitive to varietal characteristics. The pedal thresher is designed for use with short stiff-strawed (dwarf) varieties which would otherwise be double threshed by hand beating and bullock-treading. Therefore in Table 5.2, labour use patterns are given for three situations; traditional threshing of long-strawed paddy and of dwarf HYV paddy and pedal threshing of dwarf HYV paddy.9/ It is useful, first of all, to discuss the changes associated with dwarf HYVs under traditional threshing practices. In describing these changes the intention is to show that patterns of labour use have already altered as a consequence of the introduction of HYVs and to explain these changes in ways which will inform the subsequent analysis of changes in labour use associated with the pedal thresher; whilst this next section therefore is something of a diversion from our principal concern with pedal threshers we believe it provides an important perspective on pedal thresher use and it also provides the opportunity to introduce a discussion on farm management constraints which we believe are a significant explanatory factor of pedal thresher adoption.

The Effect Of HYVs On Traditional Threshing Practices

A comparison of the two traditional threshing cases (columns A and B) shows there is an increase of 68 per cent in labour costs from Tk 25 to Tk 42 per ton (row 8) with the dwarf HYVs. This increase is a product of the proportionate changes in hired labour percentage, total time per ton and the wage rate (rows 5, 6 and 7). Increases in the wage rate and percentage of hired labour considerably outweigh the reduced total labour requirements per unit throughput.

The fall in labour use is a result of the more uniform straw length and the higher grain to straw weight ratio which together make threshing easier and reduce labour requirements by nine hours per ton.

The increase in the wage rate, from Tk 1.4 to Tk 2.1 is probably due, largely, to the intensity of labour demand at the time of the boro harvest, to which the dwarf paddy results relate, compared to the aman harvest, which was the basis of column A results. For the HYV boro harvest, both aus harvesting and aman land preparation are competing sources of labour demand, but at the time of aman threshing there are fewer other large labour demands. This demand pattern is regionally specific and is relatively new in that, as discussed above, it is only consequent to the extension of the HYV boro acreage through the availability of irrigation that peak labour earnings have shifted from the aman to the boro/aus harvest. An important contributing factor is the relatively greater importance of timeliness in the boro/aus harvest. Because of rain, both the time

Table 5.2

Labour Use and Labour Earnings in Threshing

Item	Traditional A Long-strawed paddy	Threshing B Stiff-strawed HYVs	Pedal Threshing C Stiff-strawed HYVs
1. Sample size	452	30	20
2. Tons threshed	208.3	18.7	1.01
3. Total labour hours	13,040	1,005	36.4
4. Hired labour hours	3,675	376	20
5. Percentage hour hired	28	37	56
6. Labour hours per ton	63	54	36
7. Wage rate (Tk per hour)	1.4	2.1	2.3
8. Labour earnings per ton (Tk)	25	42	50[a]
9. Cost per ton if only hired labour used (Tk)	88	114	75[b]
10. Family labour hours per ton	45	33.6	16.2

Notes: a. Unlike columns A and B, the labour earnings per ton of pedal threshing (row 8) cannot be calculated from rows 5,6, and 7 since labour hours per ton (row 6) are based on total labour requirements whereas earnings per hour depend upon operator time per ton. Only 0.92 tons of the total 1.01 tons actually involved use of the twenty hours hired labour. The wage rate, based upon two-thirds of the 2.5 per cent piece rate was calculated as follows : 0.92 (tons) x 0.025 (% piece rate) x 0.66 (labour share) x 1000 (kilograms) x 3 (paddy price per kg.) divided by 20 (total hours hired).

b. Costs per ton include machine rental (one third of the piece rate received). For pedal threshing the row 9 result is the actual payment wherever the machine is hired not owned. It is assumed that the labour employed to feed paddy to the machine operators is family labour not wage labour.

Source: IDS Postharvest Project Survey.

available and the capacity to absorb delays without crop deterioration are reduced. The strength of this shift in the labour cycle obviously varies by region but, as detailed in Section 5.4, it is sufficiently strong to be very evident in the limited national statistics available on wage rates.

An increase (by nine per cent here) in the percentage of hired labour is a familiar and predictable consequence of land augmenting technical change (HYVs, fertilisers and irrigation) since family labour is in relatively fixed supply. This factor, together with the improvement in labour productivity, helps explain a reduction of over 11 hours (25 per cent) in family labour time per ton.

All three of these effects (higher labour productivity, enhanced wage rates and a proportionate increase in hired labour use) are well-known and standard results of the introduction of HYVs (see for example the review of Indian evidence by Dasgupta 1977). The net effect is an increase in paid-out costs per unit output for threshing. Such increases occur quite commonly with threshing and with other inputs for HYV cultivation compared to traditional varieties (Dasgupta 1977); the decline in net return per unit output is more than compensated by the increase in land productivity and, therefore, overall profitability measured by net returns per acre.

A factor which has so far not been taken into account in assessing the changed threshing costs associated with HYVs is the reduction, on average by 11 hours per ton, in family labour use; with a relatively fixed family labour supply the reduction in the use of family labour per unit throughput is simply a consequence of the large yield increase due to HYVs and the resultant spreading of family labour inputs.

The increase in per acre yield from adopting HYVs varies enormously by season averaging 74 per cent in aman, 160 per cent in aus and only 70 per cent in boro (for which local yield is anyway nearly double aman and aus - IBRD, 1983, p.154). The national average yield increase per acre in boro is from 0.62 to 1.06 tons and, allowing for the difference in labour productivity, requires an extra 18 hours per acre of threshing labour. At the local variety family labour participation rate of 71 per cent an extra 13 hours of family labour would be needed. This is a very similar quantity (see Table 5.2) to the reduction in family labour use per ton after the switch to HYVs which suggests that a physical supply constraint is binding (data on total family labour in threshing, before and after adoption of

HYVs, would be needed to pursue this issue more precisely but that data is not available).

At the market wage rate the imputed value of this family labour saving is significantly larger than the additional paid-out cost per ton. However, in Bangladesh, as elsewhere in South Asia, cost of cultivation studies usually suggest that market wages overstate the opportunity costs of family labour "since many farmers had no alternative use for their own labour" (GOPRB 1979, Vol.III, p.16). In the circumstances with which we are concerned here we believe this conventional assumption may be inappropriate. Is it in fact likely that poor cash-short farmers sat back and watched hired labourers do a share of the work that they used to do? Or is it likely that, at a time when demand for hired labour has pushed wage rates up, there is no other productive work for displaced family labour? The answer to both these questions must surely be no. An increase in the required volume and intensity of labour effort has caused wage rate increases and an increase in the hired share, but this is quite consistent, given yield increases, with an increase in productive family labour hours as well, if they are available. In fact, it would be very surprising if there was a fall in productive family labour time under these conditions. There may be other reasons for arguing that family labour still has a lower cost than wage labour, for example because search and supervision costs are lower; however, the general presumption that the marginal product of family labour is less than the wage rate must be qualified to take into account the seasonal pattern of labour use in agriculture (Sen 1975, chapter 8).

In principle, farmers could choose to thresh their larger yield by continuing with the same proportion of family labour and simply taking longer over the job. One reason why they do not is the risk of qualitative deterioration leading to food losses whilst the unthreshed paddy is stacked; as Table 4.7 showed extended stacking is the single most important wet season problem identified by farmers. A second reason may also be important - the farm management function of family labour. We discuss this issue next because it provides a very relevant context for the following section concerned with farmer motivation for adopting the pedal thresher.

Farm Management: The conclusion of the food losses study (Section 4.10) suggested that farm management constraints could be a useful avenue to explore in future research on food losses and technical change. 'Constraints' on farm management refer to the efficiency with which farm decision takers utilise farm resources when faced with conflicting uses of those resources. They are much easier to describe than to measure, since farm management is an input difficult to specify.

Production function approaches are not usually able to incorporate farm management as an input chiefly because a quantifiable scale or ranking of farm management cannot be physically estimated or obtained in simple interviews (see Heady and Dillon 1961, pp. 223-226). It can be expected,10/ that farm management inputs generally increase less than proportionately with scale of production (total acres or yield per acre) and reported results of increasing returns to other inputs, notably capital, may therefore be over-estimates. Unfortunately, it is in the nature of the farm management function that, alone or in combination, a set of related indices (such as education level or experience) would be a biased and, very possibly, misleading specification. However, with the introduction of South Asian green revolution technology the sensitivity of yield to variation in the quantity and quality of farm management may actually increase over certain ranges when compared to traditional agriculture. There may be a critical minimum management input which is larger than traditionally applied.11/

Notwithstanding measurement problems, it appears plausible to hypothesise that in the Bangladesh *boro* and *aus* postharvest season farm management is an increasingly important input as both the total acreage and per acre wet season yields grow. More specifically, family labour time, the instrument of farm management, becomes increasingly valuable. How much more valuable is not precisely known. Farmers that now spend less time per ton in threshing have other plots to harvest, paddy to dry and land preparation to complete for the next season; they have debts to pay and credit to arrange, inputs to buy and services to contract - as a consequence of the new technology they have more of most of these things to do. The efficiency of the farm enterprise is now more sensitive to timeliness of task completion, to successful execution of transactions, to quality of supervision (especially in land preparation to which HYV yield is much more sensitive than traditional varieties but also in water control, fertilising, disease monitoring, crop cutting and some postharvest operations).

In these circumstances, it is unlikely that family farm labour has a lower opportunity cost than hired labour.

Such an analysis would help to explain why it is family labour that is retrenched rather than hired labour when labour saving opportunities emerge. It is certainly not being claimed that this is a typical response to improved labour productivity - there are many counter-examples, including rice milling in Bangladesh, - and any apparent preference to reduce family labour is normally attributable to the growing disutility of family labour effort at higher levels of income. Nevertheless, in circumstances where seasonal influences are significant it is a pattern which has been repeated in other countries also (see eg Farrington and Abeyratne 1982, on farm power in Sri Lanka).12/ Also, the more general result that a reduction in the labour intensity of a task is associated with a reduction in supervising time has been commonly reported (Stewart 1978, p. 200); this is consistent with a 'farm management' explanation of the lower family to hired labour ratio since supervision is the main task of the family labourer as farm manager during threshing.

To summarise this comparison of farm costs in threshing of traditional and HYV seeds: higher yields due to HYVs increase demand for labour overall but also lead to increased labour productivity. Family labour is reduced more than in proportion to the improvement in hourly labour output partly because of a physical supply constraint; and, partly, we suggest, because the sensitivity of yield to the farm management input of family labour has increased - the farm management function raises the opportunity cost of family labour relative to hired.13/ There is a consequent increase in hired labour earnings which is aided by an increase in wage rates. The increase in paid out costs is more than compensated by the increase in land productivity and by the use of family labour elsewhere on the farm.

The identification of family labour as farm manager suggests that in some circumstances there is a case for a re-examination of the usual assumptions employed in measuring the relative value of family and hired labour. The purpose of putting forward this analysis here is to demonstrate that a process of change in labour use is already in progress prior to the introduction of pedal threshers and their introduction is a continuation and strengthening of this process.

The Change to Pedal Threshers

As the numbers in column C of Table 5.2 show, the existence or not of benefits to the farmer as a consequence of adopting pedal threshing methods depends entirely upon how the savings of family labour are valued. Just as between columns A and B, comparison of columns B and C show that there is reduction in total labour per unit output, increases in the proportion of hired labour, in the wage rate and in labour earnings, and, a substantial decrease (from 34 to 16 hours per ton) in family labour use. Total paid out costs (which equal 2.5 per cent of ouput) increase by more than the increase in labour earnings because of machine rental and the overall increase (Tk 42 to Tk 75) in paid out costs is nearly 80 per cent. For each ton, by spending an extra Tk 33 the farmer saves over seventeen hours of family labour effort. As Table 5.1 shows, food losses actually increase yet the rapid current growth in pedal threshing use suggests very clearly farm-level recognition of substantial net benefits from this labour saving. In the four Comilla villages where survey work was conducted all the HYV crop was pedal threshed and it must be concluded that from the farmer's perspective the investment is worthwhile. However, the per acre yields associated with columns B and C are the same; they both relate to stiff-strawed HYVs. Therefore, the "yield increase" explanation of the changed participation rate of family labour between columns A and B will not suffice as the cause of the change between columns B and C.

The change in participation rate actually has a very clear explanation - the control of pedal threshers owners over choice of machine operators when they lease out. The details of this practice have been given in the previous section and it was shown that the returns of pedal threshers to the owning household are substantially increased by their ability to provide the pedal thresher and operating labour as a package. This explanation does not tell us though why farmers choose to accept this package which forces them to replace family labour by hired labour. Since they do accept it and since the main benefit derived is family labour saving it must also be the case that this saving is greater than the additional costs incurred.

This behaviour conflicts with conventional wisdom regarding the relative valuation of family labour and the unfashionable explanation of these unconventional actions, we suggest, lies in the recognition of the farm management function of family labour. The farm management function is seasonally intense but, as described in some detail earlier, is strongly correlated with the introduction of HYVs. In developed countries there would be little hesitation in suggesting that the opportunity cost of the farm manager's time is greater than that of his hired labour - in the high-risk environment of transitional agriculture in Bangladesh, at the height of the year's activities, it is surely plausible that the farm management function places a similar premium on family labour.

Apart from the pure saving of family labour time there are two other consequences of the technical changes which facilitate farm management. First, the reduction in time spent per unit output relaxes other constraints the farmer faces such as the availability of secure stacking space and of threshing floors. There is no significant change in the intensity of labour use as a consequence of using the pedal thresher - 3 to 4 labourers are commonly used with both old and new methods; therefore the proportionate reduction in labour hours per ton reflects reasonably well the reduction in time per ton taken on the threshing floor which is nearly a forty per cent reduction. A supporting feature that, sometimes dramatically, contributes to a reduction in stacking time is that the pedal thresher can be used inside a room and therefore rain does not stop threshing: in view of the evidence of Section 4.8 on wet season losses, that qualitative deterioration is associated with extended farmyard stacking, this is an important risk-reducing feature of the new technique.

Secondly, the farmer no longer has to use bullocks (or buffaloes) in threshing his paddy. The animals are therefore available for transport and for land preparation on the farmers own land or on hire to another; they can also be rented out for threshing. No attempt has been made to include the opportunity cost of these animals in estimating costs of traditional threshing as no specific information on actual alternative uses was available; (the operation of the animal hire market is rather complex with task related rate differentials, kind payments for animal services and various socially determined non-economic considerations affecting hire rates). Needless to say, there will almost always be some positive value attached to the savings which will improve the relative profitability of pedal thresher use.

Despite the limited quantitative analysis it is reasonably clear that farmers are substantial beneficiaries from the introduction of pedal threshers. Unquantifiable justifications such as farm management are dangerous because they can be so easily abused and offered as justification for almost all forms of mechanisation. Also, there is no a priori reason why the introduction of the farm management factor to explain the profitability of private investment decision should also result in positive social returns; this will depend upon the social valuation of the old and new income streams using weights to reflect the social value of income for different levels of earnings. (See Section 5.7). From the farmer perspective though it is clear, from adoption rates, that hiring-in pedal threshers is profitable, even though the main resource saved is family labour; at the very least, this behaviour is consistent with the farm management interpretation we have suggested and, quite plausibly, is principally motivated by farm management benefits.

5.6 RETURNS TO WAGE LABOUR

The results so far have shown, predictably, that both the machine owner and the farmer have gained through the introduction of the pedal thresher. It is possible, even given the same level of output, for hired labour to benefit also since the gains to the farmer are attributable to an increase in family labour productivity by a shift out of threshing into other more productive actitivies.14/ To the extent that the substitute for this family labour is hiring of more labour rather than hiring of machines the impact of the reduced total labour time per ton is compensated.

Superficial examination of results does in fact suggest that hired labour has benefited. It can be calculated from Table 5.2 that the shift to pedal threshing raises the percentage of hired labour by a bigger percentage than it cuts total demand for threshing labour. The gain is very small (36 x 0.56) - (54 x 0.37) but at least establishes that the direct impact on non-family labour hours is not negative. Labour earnings have increased by more than the increase in labour hours since the hourly wage has gone up. Again the difference is small (2.1 to 2.3 Tk per hour) but raises earnings per ton from Tk 42 to Tk 50 (see note a, Table 5.2). Reliable estimation of hourly agricultural wage rates is notoriously difficult and in this case the problem is compounded by the fact that one set of payments is in kind. The numbers must therefore be interpreted cautiously and no great reliance should be placed on an increased earning of Tk 8 per ton when it is based upon a difference of two decimal points in the hourly wage rate - we conclude only that there has been no direct reduction in labour earning due to pedal threshers.

Indirect reductions could have occurred if the displaced family labour in turn displaces hired labour in non-threshing work. No evidence on this point is available. The essence of the farm management argument is that the release of family labour allows more timely and more intense farm management rather than mere transfer to other labouring activities; to the extent that this results in greater cropping intensity or in higher yields it will create a demand for more hired labour. To test this possibility would require a rather detailed farm management study and, given the numerous other influences on cropped area and yield levels and the problems of quantitative expression of farm management inputs, it is doubtful whether clear results could be obtained. It is not possible therefore to draw firm conclusions about indirect effects of family labour reductions in threshing but it should be emphasised that these savings are very region and time-specific; they occur at the busiest time of the crop calendar in regions where yields are increasing. These are conditions under which labour benefits are most likely to emerge and wage labour unemployment is unusual.

The use of the term 'non-family' rather than 'hired' labour earlier was deliberate, to draw attention to the changed composition of labour class in threshing. The non-family labour is that provided by the pedal thresher owner and, as Section 5.4 described, most of this labour comes from the owner's family; out of 20 threshers surveyed, 18 followed the practice of the owner determining operating labour. Of the 32 operators, only nine were labourers hired by the owner, the other 23 being members of his family. This raises the question as to how labour earnings accruing to the owning household should be treated.

If the owner family labour would otherwise have been available for work as wage labour then it is reasonable to consider total labour earnings as the benefits to hired labour. Since these gross benefits are at least commensurate with previous earnings hired labourers as a class do not suffer from the introduction of pedal threshers. Alternatively, if the owning family labourer only works for others as a pedal thresher operator and would not otherwise seek wage labour work then only 28 per cent (9 workers out of 32) of the benefits accruing to labour can be counted as benefiting the wage labour class; in this case wage labour loses since earnings per ton from pedal threshing are only Tk 14 as compared to Tk 42 from traditional threshing.

The evidence on this point is indirect. Since the availability or not of pedal thresher owner labour for the wage labour market is not known, probable availability can be guessed by looking at the economic background of the owner family as indicated by land holding. Excluding one pedal thresher owned by a cooperative and one owned by a very large farmer where accurate information on farm size was not obtainable, the average land area owned by the remaining eighteen owners in the survey was 1.4 acres. Fourteen out of eighteen owned less than two acres including five farmers who owned one third of an acre or less and two landless households; the average area owned by the eleven other farmers was just over 2 acres. Whilst these land holdings may appear small, in Chandina, Comilla, a 2 acre farmer is better off than most of his neighbours.15/ However, none of the households owning pedal threshers and reporting land ownership claimed to own more than five acres and only two claimed to own more than 3.5 acres.

So far as wage labour work is concerned, the seven poorer owners would certainly be regular participants in the wage labour market; analysis of the village survey data shows that for households owning crop land but less than half an acre of it thirty-five per cent of gross income came from labour work. For the other eleven households, participation would depend principally upon family size; the larger the number of family workers the more probable is participation. In the 1.5 to 2.5 acre farm size group in Comilla 25 per cent of gross income was non-crop income but only 5 per cent was labour income.

Overall then the results for hired labour remain slightly ambiguous. 27 of the 32 labour operators came from households owning less than 2 acres. If all of these operators are regularly seeking wage employment then the returns to labour remain unchanged as a consequence of adopting the pedal thresher. The evidence is not conclusive but suggests that overall effects on earnings are marginal. Certainly, there is no evidence to suggest that this form of mechanisation is disastrous for wage labour despite the substantial reduction in total labour per ton. Nor, given the poor to middle income status of many of the owning households (at least in Comilla) is the control over the supply of machine operators by the machine owning households a source of major concern about income distribution consequences. Indeed, unlike other patterns of mechanisation, including rice milling in Bangladesh (Chapter

Six), the low capital cost of the pedal thresher provides an opportunity, through custom-hire operation, for the relatively poor (though certainly not the poorest) to make an extremely profitable capital investment and at the same time enjoy the benefits of improved labour productivity.

5.7 SOCIAL COSTS AND BENEFITS

Since pedal threshers are being bought and used it is no surprise that the analysis at market prices has established their private profitability. The question of most interest for policy is whether this result holds good when the criterion of social profitability is applied. As Chapter Two described, there are three types of adjustment to market prices that are needed to assess social profitability; these adjustments are for price distortions, savings sub-optimality and inequality of current consumption. The first adjusts market prices to efficiency prices; the second and third adjust efficiency prices to social prices by taking account of distribution between present and future consumption and between present consumers.

As the discussion below shows, the social profitability of pedal threshers is sensitive to only one of these adjustments - inequality of current consumption. Therefore, we adopt some short-cut methods in applying these adjustments to the results from the preceding three sections on the size and distribution at market prices of the costs and benefits of pedal threshers for the three activity groups affected - pedal thresher owners, farmers hiring-in pedal thresher services and agricultural wage labourers. A full development of accounting prices is reserved for the next chapter on rice milling; in this chapter we only use plausible boundary values for the different accounting prices involved and where appropriate, estimates derived from the SBCA presented in Chapter Six.[16]

There are two reasons for reserving a fuller development of SBCA to the rice milling case. First, even the limited data available on rice milling is much better than the data on pedal threshers which is largely derived from a one-off field survey of only 20 machines. Specifically, income distribution data relating to pedal thresher use is not of a form or quality which allows reasonably clear identification of income groups. Secondly, the aggregate effects on consumption of the rice mill technology are much larger and more complex then the pedal thresher case and therefore, in a policy sense are much more important. Nevertheless, the introduction of pedal threshers is of considerable interest in understanding the relationship between labour and capital use in a transitional agricultural sector. The discussion deals first with investment and labour costs at efficiency prices. We then look at income groups and the use of income distribution weights to revalue the new distribution of income consequent to the introduction of pedal threshers.

Investment And Labour Costs At Efficiency Prices

The relative investment and labour costs of the two methods of threshing are estimated, in usual choice of techniques fashion, by comparing costs in traditional threshing and pedal threshing at a given level of output. The relevant traditional threshing results are those, given in column B of Table 5.2, for hand-beating followed by bullock-treading of stiff-strawed HYVs. Profitability at efficiency prices requires the costs of the additional capital investment in the pedal thresher to be less than the social value of the labour saved as measured by social opportunity cost.

The total investment cost at market prices of the pedal thresher was just over Tk 1000 and to estimate the investment cost per unit of throughput requires knowledge of lifetime throughput. As discussed (Section 5.4), both the years of life and the quantity threshed per year in pedal threshers are uncertain. Fortunately, a sensitivity analysis of these two parameters together with the efficiency price of the total investment and the discount rate suggests that overall social profitability is not very closly dependent upon differences in the estimated values assumed. As Table 5.3 shows, even under extreme assumptions,17/ the investment cost per ton at efficiency prices is only Tk 14. Since the change to pedal threshing results in a labour saving of 18 hours per ton the shadow wage rate would have to be low, below Tk 0.8 per hour when market wage rates are over Tk 2, to make use of the new thresher unprofitable. In Chapter Six (Section 6.9) the conversion factor from market to efficiency wages for agricultural labour is estimated to be 0.82; using this value would yield a labour saving at efficiency prices of Tk 31 and, therefore, a net social profit of Tk 17 per ton. Moreover, no account has been taken of the costs of bullock-power in traditional threshing or of the fact that the accounting price for private investment (see Appendix

6.1) is almost certainly lower than uncommitted social income: both of these factors, if included, would reduce the social costs of pedal thresher investment.

Table 5.3

The Pedal Thresher: Unit Throughput Investment Costs

	Minimum	Maximum
Outlay at Accounting Prices	1000	1500
Life (years)	15	5
Annual Throughput (tons)	66	40
Discount Rate	10	25
Investment Cost Per Ton (Tk)	2	14

Source: Text.

It is apparent that if labour inputs are valued equally then the change from traditional to pedal threshing is very profitable socially except at improbably low shadow wage rates. However, there are three types of labour - family, wage and machine owner-operator - affected by the change in technique and it is possible that the social opportunity cost, of their labour time will differ. Table 5.4 gives the share of these three labour types in traditional threshing and using the pedal threshers.

Table 5.4

Labour Share by Threshing Method (hours per ton)

	Traditional	Pedal
Wage	20	6
Family	34	16
Machine Owner-Operator	0	14

Source: Table 5.2 and text.

It can be seen from these differing labour distributions that to make the pedal thresher unprofitable at efficiency prices would require either an extremely low opportunity cost for family labour or large differences between the shadow wages of machine owners and family/wage labour.

If family labour has zero opportunity costs the social returns from pedal threshing are likely to be negative since total (undifferentiated) hired labour is unchanged but investment costs have of course increased. For the 'farm management' reasons presented earlier the private opportunity cost of family labour is well above zero and possibly higher than for wage labour. Therefore unless value enhancing distortions in the 'farm management' market are greater than in the wage labour market the social opportunity cost of family labour will not be very different to wage labour. The relative size of these distortions is not known and whilst it could be possible to develop imaginative arguments one way or the other, no great clarity can be expected.

If the opportunity cost of family labour is in fact around the same value as hired labour then (ignoring distributional issues) the pedal thresher will only be socially unprofitable if the opportunity cost of the owner-operator hired labour (14 hours) is greater (together with investment costs) than that of 14 hours normal hired labour and 18 hours of family labour. This is obviously very unlikely - the market wage rates of hired and machine owner-operator labour are the same. And, given the information on machine ownership and land holdings, there are no grounds for thinking that such large differences exist.

Distributional Issues - Analysis By Income Groups

At efficiency prices the probable level of the shadow wages of the three different labour activity groups suggests clearly that social efficiency is well served by the pedal thresher. On the assumptions that a single shadow wage can be applied to all three groups, that the market to efficiency wage conversion factor is 0.82 (derived in Section 6.9) and that the investment cost is Tk 14 per ton (Table 5.3) the pedal thresher saves Tk 17 per ton of paddy threshed. However as a consequence of the change in technique there is a change in income distribution; farmers and machine owners have gained and wage labourers have lost. The efficiency prices valuation of Tk 17 per ton implicitly accepts that consumption is equally valuable as savings (uncommitted social income) and that a unit of consumption is equally valuable no matter to whom it accrues. The incorporation of these two distributional issues, through a savings premium and consumption weights, to the result at efficiency prices will give a complete (social prices) picture of relative social profitability of the two threshing techniques.

Normally, the result at efficiency prices would be sufficient to establish that the change to the pedal thresher is also the optimal choice of technique at social prices. This is because the net effect of incorporating these two distributional issues - conventionally done through the shadow wage formula (see section 6.9) - is usually to increase the social costs of employment; since the traditional technique employs more labour the result at social prices should therefore further strengthen the case for pedal threshers. However, this usual result occurs because the premium on savings with respect to consumption at the average level, is estimated to be larger than the premium on the consumption of any particular consumption group relative to the average consumption level. As Chapter Six demonstrates these conditions do not always hold in Bangladesh. It is shown there that some consumption groups involved in the choice of rice milling techniques are so poor that increments to their consumption have a higher weight than uncommitted social income.

It is necessary to check therefore: a) Whether such a net premium on additional consumption also holds for any of the consumption groups in the threshing choice of techniques; b) whether any net loss of aggregate consumption occurs - this could only be the case if the consumption losses suffered by wage labourers have more social weight than the consumption gains by farmers and machine owners; and c) whether any such aggregate consumption loss is sufficiently large to outweigh the gain of Tk 17 per ton at efficiency prices thereby leading to a reversal of the social profitability ranking of the two techniques.

This 'checking' of the social profitability is detailed in Appendix 5.2. It is shown there that, for our field survey results, there is a net consumption gain of Tk 42 by machine owners and farmers and a net loss of Tk 12 in the consumption of wage labourers for each ton of paddy that is pedal threshed. Using the most extreme (and, frankly, unrealistic) assumptions it is also shown in Appendix 5.2 that under no circumstances could the incorporation of the two distributional weights result in a reversal of the efficiency price results by way of a higher aggregate social profit associated with traditional threshing techniques. It must be emphasised however, that this result relates very specifically to the twenty pedal threshers surveyed in Comilla during field work. We believe that these results are reasonably representative of average use patterns but a longer survey and more precise knowledge of the income levels of the gainers and the losers would be necessary to confirm this belief.

5.8 SOME CONCLUSIONS: PEDAL THRESHERS, THE EPITOME OF APPROPRIATENESS?

In many respects the evidence on the pedal thresher suggests that it is one of those rarely encountered techniques which satisfies almost every respect of 'appropriateness'. These include: simplicity in manufacture, marketing, operations and repair, and low cost, small-scale local manufacture. The rural growth linkages (the multiplier effects) of new agricultural technology are an important aspect of technology assessment that SBCA does not deal with. The importance of these linkages has been established (see Mellor, 1976) but attempts to use input-output analysis to measure 'Project' (e.g. new technology) effects (for example, Bell et.al 1982) face serious data problems. In the case of pedal threshers in Bangladesh the import content in manufacturing is low by Bangladesh standards and the backward linkages from local manufacture are therefore of potential importance. The inclusion of local growth linkages as a policy consideration would certainly work in favour of pedal threshers to a much greater extent than other forms of mechanisation occurring in Bangladesh such as the power tiller and huller rice mill. The pedal thresher is also robust, mobile, risk-reducing, and above all, a profitable investment for the owner. It displaces some labour, but is not heavily labour-displacing relative to traditional methods; since the pedal thresher can only be used effectively on shorter stiff-strawed varieties, the seasonal impact of labour displacement is restricted to those places and times (the HYV harvest mainly) where labour demands are strong and wage rates are at their highest. Certainly, a technique with these characteristics used to produce a new product in Bangladesh would not be easy to criticise on socio-economic grounds; the pedal thresher case is not so clear-cut because of the uncertain income distribution consequences, limited though they are. Moreover, farmer questionnaire results (Section 4.8) showed that delayed threshing was, almost universally, the major postharvest headache, - by improving farmer control over threshing (reduced total time and improved timeliness because rain does not stop threshing) the pedal thresher does meet a need expressed by poor as well as rich.

Perhaps the greatest virtue of the pedal thresher does not lie in its own characteristics but in the fact that successful adoption of it now makes it considerably more difficult to promote more investment-intensive threshing methods which might otherwise have been introduced. Power driven threshers are not yet widely available in Bangladesh, but if they become so they will represent a far greater threat to employment and equity objectives because of their very much more serious labour displacing effects. Raj Krishna (1975) has shown that in recent mechanisation of agriculture elsewhere in South Asia the biggest single cause of labour displacement is the introduction of mechanised threshing. The availability of an intermediate technology for threshing in Bangladesh will help to reduce the effectiveness of the large farmer lobby in pressuring agricultural policy makers and rural institutions (banks, extension services) to promote more heavily labour-displacing threshing technology.

NOTES

1. In the project survey only one case of a broadcast crop being pedal threshed was encountered. Measured losses were 7.3 per cent.

2. These results are not typical of threshing innovation involving engine-power and 'throw-in' threshing where losses are usually immeasurably small. Of course, the economics of using such technology in Bangladesh are likely to be very poor.

3. Recent results from the FAO-BRRI Project referred to earlier suggest this may be the case (Haque et al 1985 p.23).

4. At a wage rate of Tk 2 per hour there is a saving of Tk 36 per ton which is equal to Tk 67 per acre at a yield of 1.86 tons. Therefore, a cropped area of 3.95 acres would produce the required annual saving of Tk 265 (the annualised capital cost).

5. The average purchase price including transport for the twenty machines surveyed in detail in the field was Tk 1044.

6. As the machines get older this proportion will obviously increase. Only four of the machines surveyed were more than two years old.

7. See Stewart (1978) pp. 187-8 and references given there.

8. In one case the machine was owned by the local Cooperative society and in the other it was owned by a larger farmer who bought the machine for his own use. In both cases, the machines were available for hire at the full rental of 2.5 per cent without any labour.

9. The data on traditional threshing in Table 5.2 excludes labour for winnowing. Therefore, the results are not directly comparable with Table 3.5 which provides aggregate values for labour use in threshing and winnowing.

10. Heady and Dillon (1961) p. 224.

11. Hazell (1982) demonstrates increasing variability in Indian cereal output associated with use of HYVs; management is not the only, or even the major, candidate to explain this phenomenon but it probably plays some role.

12. The tractorisation literature generally tends to show an increase in family labour use (Binswanger 1078 p.53) because family labour is used to drive the tractor whereas hired labour is more commonly used to manage the bullocks/buffaloes which are replaced.

13. One weakness of this analysis is that it leaves open the more complex question of the possible effects on family farm women of a physical supply constraint reducing family male inputs in threshing; threshing is the only major crop operation undertaken at the farmyard (as opposed to the field) in which women do not enormously predominate. In purely family threshing operations they participate nearly as frequently (eighty per cent of the time) as men, but the incidence of family females working with hired male labour or of the hiring of female labour for threshing is very very much lower (below five per cent).

14. This shift is almost certainly associated (as discussed earlier) with productive opportunities for family labour in agriculture; uses such as leisure, education or non-agricultural production, certainly relevant in other circumstances, are therefore not responsible for this preference. Whilst it is hard to prove this point from most field data, which tend to give crop or plot rather than farm based estimates of labour use, the visual evidence of a Bangladesh village at the boro harvest is convincing.

15. On a farm-size categorisation, the two acre farmer has a larger farm than about 85 per cent of village households. More than sixty-five per cent of the 1.5 to 2.5 acre farm size group appear in the top four gross income deciles. (Source: project survey.)

16. Also, we use an annualised capital cost for pedal threshing investment which is an accounting procedure with identical results to normal SBCA methods but allows us to look at costs and benefits per ton at current prices without having to estimate discounted values every time assumptions about shadow wage rates or income distribution weights are changed. This is particularly convenient because we can use smaller numbers and can analyse at efficiency prices using "activity" groups and at social prices using "income" groups.

17. It is worth noting that we have assumed a ratio of efficiency to market prices of about 1.5 to estimate the maximum unit throughput costs whereas the overall ratio calculated for the rice mill in Section 6.9 is only 0.71. Given the lower share of labour in the pedal threshing costs - largely because there are no installation costs - it is likely that the pedal thresher ratio will be greater than the rice mill ratio but it is self-evidently an extreme assumption to use a value of 1.5.

Appendix 5.1

Detailed Estimate of Pedal Thresher Costs

No.	Item	Working	Amount		Total
1.	M.S. Angle:- 13'-0 =" = 22 lbs @ Tk 5.75 p.lb. = Tk.		126.50		
2.	M.S. Rounds: 21'-11"=20 lbs. @ Tk 5/-p.lb. = Tk		100.00		
3.	24-Gauge iron Sheet for both side and bottom:			= Tk	45.00
4.	22-Gauge iron Sheet for making side drums:			= Tk	38.75
5.	Ball bearings No. 6293 = 5 Nos. @ Tk 15/- each:			= Tk	75.00
6.	Bolts and nuts - 1½ kg. Tk. 35/- per kg:			= Tk	52.50
7.	Steel wire for hooks: 3 lbs Tk 8/- per lb.			= Tk	24.00
8.	Welding 2 Rft. @ Tk. 7/- p.Rft.			= Tk	14.00
9.	Painting cost			= Tk	15.00
10.	Cast iron moulding parts:	4.50			
	25 lbs. @ Tk. 4.50 p.lb. =	112.50			
	Add: Machining charges =	50.00			
		162.50		= Tk	162.50
11.	Wooden frame (Jam wood or Garjan)			= Tk	95.00
	COST OF MATERIALS			Tk	748.25
12.	Direct labour:				
	(a) Mechanical section = Tk	140.00			
	(b) Wood and carpentry = Tk	25.00			
		165.00		= Tk	165.00
	TOTAL PRODUCTION COST			Tk	913.25
13.	Profit and overhead				
	(a) Profit =	5%			
	(b) Overhead =	10%			
		15% on Tk 748.25		= Tk	112.25
	TOTAL			= Tk	1,025.50

(M.S. = mild steel, Rft = rivet feet)

Source: M.A. Mansur (Accountant) and M.D. Khurshed Alam (Asst. Production Manager), the Commilla Co-operative Karhana Ltd, Ranir Bazar, Comilla.

Appendix 5.2

Distributional Issues and the Choice of Threshing Technique

This appendix is concerned with the issue of whether the use of distributional weights - a savings premium and weighting of current consumption - could make traditional threshing a more socially profitable technique than the pedal thresher which has been shown to be the more profitable technique at efficiency prices. The first requirement is to use the survey results to identify the effects of the change in technique on consumption.

The farm family hiring-in the pedal thresher saves nearly eighteen hours of labour for which the imputed value at market wage rates (Tk 2.1 per hour) is just Tk 4 more than the additional paid out costs of Tk 33 (see table 5.2) for machine rental and hired operating labour - a consumption gain of Tk 4.

Machine-owners receive a machine rental of 8.3 kilograms (one-third of the 2½ per cent kind payment) per ton with a value of Tk 25. Also, as Section 5.4 described, out of 18 machines surveyed machine-owners supply thirty-two operators (in four cases there is one operator and in fourteen cases there are two) of which 23, or 72 per cent, were family members. Thus, they receive 72 per cent of the payment to labour which is (0.72 x Tk 50) Tk 36; in total they receive Tk 61 per ton. The costs of machine ownership have to be deducted from this income. These costs, annualised over the life-time of the machine, were estimated (Section 5.4) as Tk 265 per year. With the minimum estimated throughput of 40 tons per year the cost is Tk 6.6 leaving a net income of approximately Tk 54 per ton.

Hired labour has lost its earnings (Tk 42 per ton) from traditional threshing and obtains 28 per cent of the earnings for operating labour on the pedal thresher equal to (0.28 x 50) Tk 14. Wage labour losses total Tk 28.

Overall the changes in income amount to a net private gain to consumers per ton of Tk 30. Farms hiring-in, (Tk 4+); machine-owners, (Tk 54+); and wage labourers, (Tk 28-). To apply income distribution weights these activity groups have to be reorganized into income groups. As Section 5.6 described, the limited information available on the economic characteristics of pedal thresher owners suggests that at least seven pedal thresher owners, who are either landless or own less than half an acre of land, are regular participants in the wage-labour market.

These seven include four machine operators who work alone and three who work with two family members. Therefore the rental income from seven out of the eighteen machines and the wages earned by ten of the twenty-three owner-operators are properly considerd as hired labour earnings. The result of this adjustment is that machine owners income is reduced by Tk 16 to Tk 38 per ton and wage labourers income is increased by the same amount leaving a net reduction in wage labour earnings of only Tk 12. Thus, by income groups, the total gain of Tk 30 consists of gains of Tk 4 for farms hiring-in and Tk 38 for machine owners and a loss of Tk 12 for wage labourers.

With no data available on the relative income levels of these three groups the analysis has to proceed by indirect methods. The simplest case to examine is the grouping of farmers hiring-in and machine owners in a single income group with a gain of Tk 42 compared to the loss of Tk 12 by wage labourers. It is evident from these figures that the weight on consumption for the wage labour group would have to be considerably greater than unity if the Tk 12 lost by them is to be considered more valuable than the gain of Tk 42 by the beneficiaries and the efficiency price gain of Tk 17. In fact using the values from Appendix 6.1 it is possible to demonstrate that under no realistic assumptions about consumption distribution weights could the consumption losses by wage labourers result in a reduction in aggregate social welfare by adoption of the pedal thresher.

Assuming that all income is consumed and using the highest weight for any of the consumption groups identified in Appendix 6.1 of 2.5 the loss in consumption by wage labourers is Tk 12 x 2.5 = Tk 30. This weight is that used for the households of female wage labourers who are amongst the poorest households in Bangladesh; it is a composite weight of 2.85 as the premium on these households consumption relative to the average level and 1.14 as the premium on savings (uncommitted social income) relative to consumption. The consumption conversion factor (see Section 6.9) is estimated to be 1.08 so the gain in consumption at accounting prices is Tk 30 x 1.08 = Tk 32.4.

On the assumption, temporarily, that the investment in pedal thresher is made from uncommitted social income then the net weight on the consumption gains by farmers and machine owners would have to be 0.34 in order to equalise the social profitability of the two techniques. (The social value of this Tk 42 consumption has to be equal to Tk 32.4 - Tk 17 = Tk 15.4 at accounting prices or Tk 14.26 at market prices. Dividing Tk 14.26 by Tk 42 gives a weight of 0.34).

The weight of 2.5 on wage labourers was based on a value for the elasticity of the marginal utility of consumption (see Appendix 6.1) of 2, a critical consumption level of Tk 1679 and an average consumption level of Tk 1791. The weight of 0.34 on farmers and machine owners consumption implies that they have a consumption level 72 per cent above the critical consumption level and 61 per cent above the average consumption level.

Many of the farmers using the pedal thresher will actually be quite poor so these income levels are unlikely but even if they were true for machine owners (who are the major beneficiaries on the assumptions made) the resultant low weight on their consumption has another implication which critically affects the results. The temporary assumption was made above that the investment in pedal threshers is made out of uncommitted social income. In fact the investment is made out of funds available to the investing household some part of which will quite probably be available as a consequence of sacrificing consumption. The cost of funds available by sacrificing consumption is equal to their market value (at accounting prices) multiplied by the consumption weight. For example, if all of the Tk 14 per ton costs of the pedal thresher investment (see Table 5.3) were actually a result of foregoing consumption, the investment costs per ton are only Tk 14 x 0.34 = Tk 4.76.

In practice, some investment funds will be from consumption funds and some from savings but if the social value of those funds is less than uncommitted social income then a downward adjustment in investment costs must be made. Little and Mirrlees (1974 pp 201-2) suggest the social value of private investment funds will always be less than uncommitted social income. Some downward adjustment on investment costs will therefore always be required and consequently, even with the extreme assumptions that are made here concerning income distribution weights, the pedal thresher will remain the socially optimal choice of technique given the consumption distribution effects indicated by the field survey.

In evaluating this result the first thing to stress is that it is based on a limited survey and a number of simplifying assumptions; it cannot be considered a robust result but at least provides some evidence of how the incorporation of distributional issues could inform a policy view of the change in threshing technique. To provide a clear policy interpretation requires a larger sample size and a knowledge of the actual income levels; critical here are the income levels of machine owners and labourers rather than farmers hiring-in since these farmers have only a small income gain. Aggregate data on income distribution patterns from the project survey villages does provide some help; data is available on the size-class distribution of land owned, cross-tabulated by decile groups of gross income per consumption unit. Excluding the poorest seven owners who were included in the hired labour income group, the pedal thresher owners, on average, have just over 2 acres of land and so the size-class, 1.5 to 2.5 acres, is reasonable to use as an approximation of their farm-size ranking. This size class is distributed very unevenly across the decile groups with, for example, over fifty per cent of cases in the top three deciles and under eight per cent in the bottom three deciles. This data suggests that the differences in income between the two groups may be very large but must be contrasted with the fact that there are another seven owners (out of twenty machines surveyed) who are landless or own less than half an acre of land.

In conclusion, it is worth emphasising two more general considerations that condition interpretation of the social profitability of the change in technique. First benefit-cost methods are not sensitive to loss of livelihoods in that they take no account of the hardship associated with complete removal of earnings opportunities; low shadow wages and high distributional weights by themselves do not carry any guarantee that alternative earnings in that place and that season are actually available (the next two chapters provides an illustration of this). The way in which pedal threshers are seasonally selective in their displacement effects is a major advantage therefore. The pedal thresher is used at a time of peak labour demand when alternative labour opportunities exist.

A further consideration is the low absolute level of earnings in agricultural wage labour in Bangladesh. Welfare improvements require increased labour productivity; this can only result from investment. The pedal thresher enjoys low investment-output (Tk 14 per ton processed) and investment-worker (Tk 250 per workplace) ratios and is therefore a relatively equitable type of investment. Both the wage labour operator and the machine owner-operator share the benefits of the improved labour productivity. Whilst this provides no satisfaction to those wage labourers that have been displaced the low capital cost of the thresher leaves open the possibility of sharing some of the access to the higher productivity and wage levels as pedal thresher use grows; development policies and projects which are targeted on agricultural wage labour will be able to utilise the pedal thresher as a method to raise labour productivity and self-reliance by increasing the proportion of pedal threshers owned by the poor.

6

Technical Change II: Rice Milling

6.1 INTRODUCTION

In Bangladesh traditional manual methods of transforming paddy into rice are being replaced by mechanically powered rice mills. We have observed earlier (Chapter Two) that this process of technical change is often regarded as a serious problem because of the displacement of hired female wage labour. This Chapter presents a systematic analysis of the social and economic characteristics of the process of technical change using field survey data from the IDS postharvest project. The results reported here set the context for a policy-oriented analysis of the economic status of female wage labour which is the subject of Chapter Seven.

Section 6.2 describes the traditional dheki technique, its distribution and the types of labour used; Section 6.3 provides a similar picture for the newer technique, the Engleberg huller mill. Section 6.4 compares the two techniques with respect to the quantity and quality of rice they produce per unit of paddy input. Sections 6.5 and 6.6 are concerned respectively with the farm-level decision criteria for choice of milling technique and with the private profitability of owning huller mills.

The major arguments of the chapter are in Sections 6.7 to 6.11 which are devoted to a social benefit-cost analysis of the choice of techniques. After an outline of the approach (6.7), the relative profitability of the two techniques is compared at market (6.8), efficiency (6.9) and social prices (6.10). The analysis shows that at either market or efficiency prices the huller mill is much more profitable but that as soon as distributional issues are incorporated through the use of social prices, relative social profitability is reversed. Section 6.11 is a sensitivity analysis which shows that this result is robust and the social profitability of the *dheki* is much greater than that of the huller mill even with substantial changes in the parameters and weights used.

6.2 TRADITIONAL PROCESSING: DESCRIPTION, DISTRIBUTION AND TYPES OF LABOUR USED

In addition to the *dheki* there are two other traditional methods of husking rice in Bangladesh. In the east of the country, particularly Sylhet district, the most common practice is to use a hand-held wooden pestle and a mortar identical to the type used elsewhere in Asia and in Africa. In the south-east tip of the country, southern Chittagong district, the *dolong*, sometimes called the *Teknaf dheki*, is used for the first stage (husk removal) and the *dheki* is used for the second stage (polishing). The *dolong* is an intricate wood and sun-dried mud construction that husks the paddy through the shearing action of a rotating arm against which the paddy is pressed as it moves from top to bottom inside the round body of the machine (Lockwood 1981, pp.3-7). Both of these techniques are used in areas where raw rather than parboiled rice is consumed which probably explains their specific geographical dispersion. The study does not deal with these two minor traditional techniques though it is likely that the arguments developed with respect to the *dheki* apply more or less to them also, since the essential economic characteristics are similar.1/

The *dheki*, as described in Chapter Three, is a foot-operated pestle and mortar. The pestle is a short wooden peg (usually with an iron band strengthening the working end) set in a narrow wooden beam of four to eight feet length; this beam is pivoted about two-thirds of the way along its length so that it can be raised (or lowered) by applying (or releasing) a downward pressure on the other end. The mortar is simply a bowl-shaped hole in the ground, lined with wood or cement, within which the paddy is placed; the paddy is dehusked and polished by the shearing action of the pestle against the paddy. In addition to the *dheki* itself there is a small frame to provide a support for the women operating the *dheki* and usually the working area is covered by a bamboo supported thatch roof.

The majority of rural households own a *dheki*. In a 1978-79 project survey of eight villages in Comilla and Tangail nearly three-fifths of the 3,094 households owned one; and, given the social structure of the Bangladesh village,2/ every household has access to a dheki. The share of national rice production which is processed by the dheki can only be calculated approximately, as the difference between total paddy production and the quantity processed by mechanised rice mills. This issue is explored in the next section on rice mills and we turn here to another critical dimension of dheki processing, the pattern of labour use.

DHEKI LABOUR USE: Processing of rice in dhekis is invariably performed by women and in most cases by family farm women. Since national data on production by farm size and on proportions of hired and family female labour by farm size are not available it is not possible to be precise about the relative involvement of hired and family female labour types.

In our survey of eight villages (3094 households) in Comilla and Tangail districts there were 2163 farms producing rice and 297 (13.7 per cent) used some non-family female labour in rice processing. These figures almost certainly underestimate the proportion of total village paddy production that is processed by hired female labour since it is richer farmers that use hired female labour more frequently; dividing households into decile groups based on gross income per consumption unit we found that 30 per cent (76 cases) of farmers in the top decile hire in compared to just 4 per cent (9 cases)3/ in the lowest decile. [A similar picture emerges using farm-size data; 60 per cent of farms above five acres hire in whereas only 4 per cent of farms below half an acre do so, but the figures require a cautious interpretation since there are 727 farms in the latter category but only 52 in the former.]

Further data from Comilla on labour use by farm for the largest plot cultivated showed that over 9 per cent of total female labour (2852 out of 30,267 hours distributed between 955 cases) was hired (each farm gave information from one season but all seasons were represented in the overall sample); but the figure of 9 per cent must underestimate the true share since it underrepresents large producers' share in total production. As a crude attempt to introduce some more accuracy we multiplied, for each decile group, their proportionate share in total area cultivated by the proportion of farms using hired female labour for the one plot per farm that was surveyed. Summed, this gave a figure of 17.3 per cent of rice area cultivated by farmers who hire in female labour for dheki work; to be regarded as an accurate estimate of hired labour share one would have to assume that hirers in always hired in, that non-hirers never hired in and that yields per acre were constant.

These assumptions are untested but field visits suggest that the pattern of *dheki* labour use does have a seasonal aspect. Rice processing is an activity that is associated with rice consumption rather than rice production - paddy can be stored and husked when needed. As such, one could expect to find a fairly stable seasonal pattern of labour use - at least amongst those farm families who can feed themselves; but in fact, *dheki* work increases at harvest time not because there is *more* paddy available but because there is more rice to be cooked for the male labour hired to assist in cutting and postharvest operations. One consequence of this is that some farm families have female labour to help with the harvest-time *dheki* workload but they do not do so at other times. Therefore, given our method of estimation, 17.3 per cent currently overstates the proportion of female *dheki* workers that receive wages.

However, there is also an upward correction to be made to the hired labour share in *dheki* work for rice farmers that use the rice mill. 17.3 per cent of rice farmers hired in female labour but, as described more below, 19.0 per cent of farmers in the same Comilla survey, with 26.8 per cent of total rice area, were rice mill users. Several of these cases will also have been on a seasonal basis only but some upward adjustment of the estimate is still required. On the strong assumption[4/] that the 19 per cent rice mill users never used the dheki and therefore never needed to hire female labour, the percentage of hired female labour in rice husking amongst *current dheki using* farms would be about 22 per cent of farms cultivating about 24 per cent of the area. Thus, there is one reason for lowering the estimated share of hired female labour in dheki work, by what amount is not clear, and one reason for raising it, by about 7 per cent to 24 per cent. Arbitrarily, we have chosen to make neither of these counterbalancing adjustments here but the sensitivity analysis (6.11) considers the implications of making such adjustments. Despite the inability to fine-tune these figures, they give fairly clear evidence that family female labour predominates and probably was responsible for 80 to 90 per cent of total traditional processing at the time and place of the survey. Jabber (1982) found the share of family labour to be 80 per cent.

The final caveat, 'at the time and place of the survey' is of some importance, since, as mentioned, at that time (last quarter 1978) 19 per cent of farmers in these four villages, with 24 per cent of surveyed production, were already using huller rice mills. There is evidence to suggest that the first farmers to start using the mill are those that save payments to labour rather than those who stand only to save family female labour. As a consequence, there is a fall in the proportion of the _dheki_ labour force which is hired as substitution of _dhekis_ by rice mills occurs.

The first evidence on this point is from a 1980 project survey of 100 farmers in the project villages in Tangail; mills were introduced in the area during the period of fieldwork and the survey showed that one-third of farmers started using the rice mill within a year and that 75 per cent of these mill users had previously used hired female labour; they accounted for nearly two-thirds of all farms that ever hired female labour. Secondly, Ahmed (1982) in a survey in Mymensingh district found that with the introduction of rice mills the ratio of hired labour displaced to family labour displaced was 16:13.

The explanation is that in shifting from _dheki_ to mill the costs saved by reducing use of hired female labour are greater than the costs saved by reducing family female labour (the opposite situation to that of harvest time male labour discussed in the last chapter). Thus, in areas where the infrastructure is better developed and the density of mills is higher than average the proportion of _remaining dheki_ work completed by hired labour will be less than the national average. Specifically, it is possible that the survey results from Comilla, with a well-developed infrastructure, showing about 17 per cent of _dheki_ processing performed by hired labour underestimate the national share; because of this and the other sources of bias in this estimate, the SBCA will include different shares of hired and female labour in the sensitivity analysis (Section 6.11).

6.3 RICE MILLS: DESCRIPTION, DISTRIBUTION AND TYPES OF LABOUR USED

Rice mills in Bangladesh fall into two categories; the first, known in the literature as major mills or commercials mills, are principally used by traders or on contract to the Food Department. The second category, known in the literature as custom huskers (Tickner 1974), custom mills (GOPRB 1978b) or rural mills (IBRD 1983), are smaller mills, almost all Engleberg hullers and principally used by farmers. It is this second group which are gradually displacing dheki processing of on-farm consumption paddy, and which are the focus of this chapter; first, though, we make some brief comments on the first category and in what way it is distinguished from the second.

COMMERCIAL MILLS: In Bangladesh, farm sales of rice are rare and the farm-level surplus is usually marketed as raw paddy; the commercial mills provide most of the parboiling, drying and milling facilities for this surplus. In a recent and comprehensive review of the available evidence, IBRD (1983, pp.35-58) estimated that in 1981 about ten per cent of production was processed by 237 commercial mills (including 82 not operating). This is almost certainly an underestimate since there are a number of smaller commercial mills which are not included in official statistics but which depend mainly upon traders rather than farmers.

A mill is classified as major by the Food Department when the mill facilities provided are deemed suitable for the mill to be allowed to bid on or accept government contracts. Most of the commercial mills are major mills or large mills with aspirations to the recognized status of major mills; they are distinguished from the second category mills, used by farmers, by the services they provide and their scale of operation. These two characteristics of major mills account for the higher levels of capital investment and their higher output to labour ratios.

Major mills provide parboiling and drying facilities which farmers' mills do not; frequently, major mills also provide storage facilities, if only short-term. These are services required by the rice trade but not by farmers who parboil, dry and store at home. These commercial mills have a capacity of up to three tons of paddy an hour, more than three times as large as the biggest farmers' mill.

The investment costs for major mills varies enormously because there are several different techniques in use. The most common is a scaled-up version of the single-unit Engleberg huller used by farmers but with a battery of hullers run off a single large power source, which may be electricity, diesel or steam. There are also under-run disk hullers, rubber roll mills and fifteen fully automatic rice mills. Loans have been approved (1981) for a further 57 automatic rice mills (IBRD 1983, p.42). These automatic rice mills, with an average investment cost of Tk 7.5 million, require thirty to forty times more capital investment than farmers' mills. More generally, commercial mills have much higher capital costs per workplace resulting in lower labour use per unit of throughput, though labour use varies according to the degree of automation. Unlike farmers' mills the commercial mills employ female wage labour but only for parboiling and drying; male wage labour does the other tasks.

The economic performance of the automatic mills is extremely poor for several reasons (IBRD 1983, pp.44-45) but they only account for about a quarter of total commercial mills. The overall economic performance of the commercial mill sector is very directly influenced by government food policy especially as regards average milling ratios, paddy-rice price differentials and levels of milling charges for work on the government account. However, so far as the impact on dheki use is concerned, the commercial milling sector is not a major competitor. These mills are not generally used by farmers. Whilst the larger farmers' mill does sometimes compete with these commercial mills for the business of traders, most farmers' mills rely for most of the time on the milling of paddy for on-farm consumption. In this, they are directly competing with the dheki and the SBCA given here concerns the comparison between the Engleberg huller mill service available to farmers and farm-level dheki processing of paddy for on-farm consumption.

HULLER MILLS FOR ON-FARM CONSUMPTION RICE: The Engleberg Steel Roller Huller Mill (hereinafter called the huller or rice mill) has been described above (in Section 3.3) and a full engineering description (including results on attempts to improve the design) is given in Arboleda (1975). To recap briefly, it consists of an enclosed fluted steel roller that rotates; paddy is fed in at one end of the roller. The husk, bran and small broken grains pass through a net (part of the casing) and the husked rice passes out separately at the far end of the roller. Usually, a single pass is performed for both husking and polishing. The roller is driven by a diesel or electric motor connected by flat belts and pulleys. A preliminary survey of 35 mills in the two project areas showed that their engines/motors varied from 14 to 40 horse power. Operators claimed throughputs per hour varied from 0.22 to 0.93 tons, though their rated capacity is greater than this (Asia Foundation 1980). Differences in throughput are due in part to the different sizes of rice mill used, and in part to the age of machinery.5/

All the mills surveyed had wheat grinding facilities - 16" to 20" grinding stones. Unlike paddy processing the mills enjoy a virtual monopoly6/ on wheat processing and with wheat acreages and yields in Bangladesh growing annually (in 1983 acreage was 1.4 million and yield was just over 1 million tons, GOPRB 1983a) the grinding of wheat is becoming an increasingly important source of earnings - see Section 6.6.

As the preceding discussion suggests almost all of the marketed surplus is mill processed and mainly by commercial mills, not the farmers' mills we are concerned with here. It is very difficult to establish how much of the eighty per cent or so of the rice production that is consumed on farms is also mill processed. The Comilla data cited above showed that about 19 per cent of rice farmers with about 27 per cent of rice area used the mill. One way of establishing whether this sample result accurately reflects the national share would be through data on the number and size of farmers' mills and their capacity utilisation. Official estimates of the number and distribution of the small mills are available (cited for example in Jabber 1982) but the figures are underestimates since they only reflect the number of licensed mills. In the survey of 35 mills referred to above only 24 gave a clear answer when asked if they had a licence and in 15 cases the answer was no. (Predictably, it was found that mills on main roads were usually licensed but interior mills often were not.) The situation is further complicated because official figures of the Food Department are seriously inconsistent with estimates from other departments (see Noor 1980, p.3). The uncertainty is compounded by the tremendous regional variation, apparent from extensive travelling in the country - in Bogra dhekis seemed to have virtually disappeared in 1980 but in Tangail in 1979 there were very few mills to be seen outside of market towns - and by the sharp differences between roadside and interior villages.7/ With all these complications it is only possible to give a probable range of the share of rice mills in milling farm consumption paddy; based on some heroic assumptions,8/ this range is 25-35 per cent. This is very similar to the conclusion reached by IBRD (1983), after a review of recent estimates, of 25 to 30 per cent for the share of small rural mills in national rice processing; (the IBRD estimate assumes that the larger commercial mills have ten per cent of output so their estimate for the small mills also includes about half of the marketed surplus. But our figure assumes that the official estimate of mill numbers is seriously understated).

With the current programme of rural electrification the rate of increase in mills is likely to speed up - the Rural Electrification Board (REB) were receiving an average of 200 rice mill applications for each new thana electrified in 1980 (personal communication, the Chairman, REB, August

1980). Official figures on mill numbers indicate a growth rate of 700 per annum in the four years to 1981. Thus, while it appears that the dheki remains the most common method for processing rice for on-farm consumption, the rice mills have already made substantial inroads to the sector and are likely to continue doing so at an increasing rate.

Finally, on the types of labour used in the farmers' mills the two important characteristics to note are that; (i) unlike the dheki or mills with parboiling units, the labour force is entirely male,9/ and (ii) in the thirty-five mills surveyed the total size of the labour force ranged from 1 to 4 and averaged 2.45 men of which 1.18 were hired; i.e. the mills generate little more than one job per mill for wage labour. In addition to this mill labour, the customer (the farmer) has to transport the paddy to and rice from the mill and the by-products have to be separated by family farm women. These labour inputs are estimated to be about one and a half hours per maund (transport) and thirty minutes per maund separation of by-products - see Sections 6.4 and 6.5.

6.4 QUANTITY AND QUALITY OF MILLED RICE OUTTURN - A COMPARISON

The Engleberg huller mill has been subject to a great deal of critical comment because of its supposed poor performance. The relatively high temperatures generated and the pressure that the grain is subjected to during the milling process reduces milling yield and causes a high proportion of brokens (Arboleda 1975). However, the evidence on the poor performance of these rice mills relates to raw (unparboiled) paddy. Harriss (1979, p.36) has found, in a review of Asian evidence, that in the case of parboiled paddy the steel roller huller mill is not responsible either for lower milling ratios or high proportions of brokens, except in cases of poor management or poor quality rice. Nevertheless, in Bangladesh, the financing by local banks of more modern rice processing facilities has been posited upon a supposed two per cent increase in rice outturn by substituting for the huller mill.10/

To examine this issue a study was undertaken to compare the percentage milling yields and the percentage of broken grains in three milling techniques; the dheki, the huller mill and a laboratory-model modern mill using rubber roll hulling and stone cone polishing. These results were given in Table 4.2 and repeated here for convenience (Table 6.1).

Table 6.1
Food Losses and Technical Change in Rice Husking and Polishing

Technique	Per cent Milling Yield	Per cent Broken Grains	Sample Size
Dheki	72.02 (0.46)	28.28 (4.50)	20
Huller Mill	69.94 (0.33)	30.54 (4.52)	20
Laboratory Model Modern Milling	71.87 (0.28)	8.89 (1.91)	20

Note: Numbers in brackets are standard errors
Source: IDS Project Survey.

The methods used are reported in full in Shahriar (1980) and summarised here; twenty lots of parboiled paddy were purchased and divided into three parts which were milled using the three different methods. The weight of paddy in, the weight of rice out and the percentage by weight of broken grains (obtained using indented plates in the laboratory) was estimated in each case. The method standardises results for paddy variety, quality and moisture content.

Tests on the results established that the huller mill's lower milling yield is statistically significant (a t-test at the 95 per cent level) compared to either the dheki or the laboratory mill11/ but that the marginally better performance of the dheki over the laboratory mill is not statistically significant. On per cent broken grains, both the dheki and the huller mill produced statistically significant higher levels than the laboratory mill.

In other words, in Bangladesh the shift to the huller mill is causing a reduction in milling yield from 72 to 70 per cent which is equivalent to a loss in food availability of nearly three per cent. However, these mill-induced reductions in rice output are recovered when the by-products are separated. The farmer using the mill takes back rice and by-products which consist of husk, bran and broken rice. These are then sorted by family farm women using a winnowing tray. This work takes about half an hour a maund and, though not measured, impressions during field work suggest that more or less complete recovery of the rice mixed with the bran and husk can be achieved. (The task is the same as _dheki_ workers perform during _dheki_ processing where all of the husk, rice and bran are mixed in the mortar - usually the paddy goes through three passes in the _dheki_ and after each pass the rice, husk and bran are separated; in _dheki_ work about ten per cent of the labour time is spent in this cleaning process.) Thus, whilst the huller mill is inferior to the _dheki_ technically as measured by immediate rice outturn, in practice use of the huller does not result in less rice being available ultimately because of post-milling cleaning of by-products.

Most of this rice which is recovered from the mill by-products is in the form of broken grains which pushes the percentage of brokens up from around thirty to about thirty-two per cent (Table 6.1). _Dheki_ use results in about twenty-eight per cent brokens. These results contrast with an average measured level of brokens in modern milling methods of less than nine per cent; the following few paragraphs assess the significance of the different levels of broken rice both between _dheki_ and huller and between these methods and modern milling.

The higher percentage broken grains requires a two-stage assessment. First, some broken grains are separated from the whole rice by farm households and used in alternative, and probably lower value ways; the two most common uses are for feeding chickens and for petty barter trade with village hawkers. Now, the percentages of brokens separated in the rice cleaning process undertaken by farm-women and used for these purposes are much much lower than the measured percentages in the laboratory. In the case of the _dheki_, the average percentage of broken grains separated was just 2.3 per cent (compared to 28.3 per cent in the experimental work) and in huller mills was 4.9 per cent (compared to 30.5 per cent.) In effect, the farm women who do this work only separate the chips and small brokens leaving the large brokens to be consumed with the rice - the much higher figure (4.9 per cent) for the huller mill, though not statistically significantly different to the _dheki_ figure (2.3 per cent), indicates that the huller mill produces more of this lower value product.

Even so, in both cases the percentage of brokens separated by farm-women is low, the economic loss associated with this change in use, from food to poultry feed, is not precisely known but if it is as much as fifty per cent it implies a one per cent reduction in the value of _dheki_ output and a two and a half per cent reduction in mill output. This estimate is very arbitrary however, there was a wide sample range for the percentage of brokens and the dheki-huller differences were not statistically significant; therefore this effect of brokens on values of milling yield has been neglected in the initial SBCA comparison of dheki and huller but the sensitivity analysis looks at the effects of changing the assumptions regarding the value of output.

Secondly, given that most of the brokens are actually consumed along with the rest of the rice it is necessary to establish whether this leads to a reduction in the availability of cooked rice due to loss of solids during cooking. To answer this question the rice samples from the milling experiments were used in two sets of cooking tests; one set was the obtained rice including brokens and the other had the broken grains separated. The experimental methods and results are fully reported in Dawlatana (1980). Her results are summarised in Table 6.2.

Table 6.2
Total Solid Loss During Cooking

Sample size	Milling method	% Average Solid Loss in Cooking: with brokens	% Average Solid Loss in Cooking: without brokens	t-values for difference between means
16	mill	9.89	7.62	4.99
16	dheki	9.25	7.14	4.83
t-values for difference between means		5.9	8.9	

Note: All mean differences are statistically significant at the 1% level.
Source: Dawlatana (1980).

The results show that with both techniques there is a two per cent reduction in availability of rice for consumption at the given level of brokens - and Dawlatana (1980) shows that there was a simple linear relationship between percentage brokens and percentage losses. (This result is based on a comparison with rice containing no brokens and would only indicate a positive gain of two per cent in cooked rice using modern milling methods on the assumption, unfavourable to both dheki and huller, that no loss of solids occurs in cooking due to the nearly nine per cent of brokens associated with the laboratory-model modern milling methods.12/ These losses are caused by the greater tendency of brokens to dissolve during cooking compared to whole grains.

The differences due to the two techniques were also statistically significant but the actual physical difference, in the case of 'with brokens', was only about half a per cent. The solid content reduction is not a complete loss however because the rice water, containing the dissolved rice, can be, and generally is, used; this gruel is given to children, the old or the infirm and occasionally fed to animals.13/ It is by no means clear therefore that any nutritional loss is sustained by the difference in solids loss; the initial analysis in the SBCA given below assumes that there is no difference in value of output due to solids loss in cooking (but the effect of changes in relative value of output are considered in the sensitivity analysis).

A further issue on product differentiation by technique concerns utilisation of by-products. Both husk and bran have industrial uses, the most well-known being husk as a fuel and bran as a source of oil. It is sometimes argued, in favour of modern techniques, that the more valuable uses of by-products are not realised with small-scale milling technologies such as dhekis and huller mills and further, that the bran which is a more valuable by-product in several uses, is lost because it is mixed with the husk. By-product utilisation can certainly be improved when milling technologies similar to the laboratory-model become viable but the operation of these milling methods on rice for on-farm consumption will not be commercially viable in the immediate or medium term future because of technical rigidities in the scale of operation. Mills using these techniques are much larger than the huller mills and have much higher fixed investment costs; milling on-farm consumption rice they could only achieve economically viable capacity utilisation by raising average transportation costs between farm and mill much above the incremental value attributable to higher rice yields. They are unlikely to be cost-effective, even with their potentially more valuable by-products, in the foreseeable future in Bangladesh.

On wasteful use of bran currently, it is almost always the case after both dheki and huller milling that the bran and husk are separated. The bran is usually used to feed poultry and the husk, in dung cakes, is used as a domestic fuel. Their initial admixture may result in some bran being used with the husk but the amount is small and anyway the difference in value of the use is likely to be very low (though difficult to quantify). So far as our <u>dheki</u> - huller mill comparison is concerned, by-product utilisation does not effect the results.

The essential conclusion of this section is that the amount of food available for on-farm consumption as a consequence of a shift from the <u>dheki</u> to huller milling is unchanged though about half of one per cent is available only in the gruel rather than as whole rice because of loss of solids in cooking and a higher (though not statistically significant) percentage of brokens is separated from the whole rice. At no time in farmer (male or female) interviews was a <u>dheki</u> preference indicated because of these differences and, overall, there seemed to be little or no awareness of them. Moreover, there appeared to be no varietal-associated preference for either of the two techniques and nor was grain moisture content a preference factor since, in both techniques, paddy was milled at about fourteen per cent moisture content. The occasional statement indicating preference for <u>dheki</u> milled rice is more common in the towns than in the countryside and is probably sentimental attachment of urban and peri-urban people to their memories of rural living. Consumer preferences for specific product characteristics are anyway not obviously an important influence on the choice of milling technique (see sections 3.6 and 4.8 for discussion on consumer preferences - Ahmed (1982, p 118) suggests <u>dheki</u> preference will inhibit farmer use of the mill but gives no evidence). Overall, it appears that technique-induced qualitative differences in the rice do not strongly affect farm-level decisions about choice of technique.

6.5 THE FARM-LEVEL DECISION CRITERIA FOR CHOICE OF MILLING TECHNIQUE

In Section 6.3 we have seen that at least a quarter of all on-farm consumption rice is milled mechanically by huller mills. This section tries to determine the economic rationale for farm decisions to adopt or not to adopt the more recently available technique.

In the most common case where traditional <u>dheki</u> processing is performed entirely by family female labour the situation is reasonably uncomplicated to analyse.

Essentially, the trade-off is between the *dheki* investment and a large number of family female labour hours as opposed to a (much) smaller number of family or hired male labour hours and a small cash payment. Before examining the labour components of the trade-off more closely it is useful to deal with the capital investment associated with traditional processing, the investment in a *dheki*.

Estimates of the cost of a *dheki* vary widely. In bank loan schemes in 1980[14] a purchase cost of Tk 350 was assumed and IBRD (1983, p.51) use a value of Tk 400-600. Other recent estimates based on field surveys include Tk 120-130 with up to thirty years life (Harriss 1979) and Tk 209 with fifteen years life (Jabber, 1982). The bank loan scheme estimate was not expected to be an exact estimate and the IBRD estimate is from an unknown source; however, both of these estimates concern large *dhekis* used by women working in the paddy husking business and certainly overstate actual farm-level costs. The differences between the two field surveys, whilst in part perhaps related to sample sizes (9 and 80 respectively), are almost certainly most heavily influenced by regional differences between the districts involved in the surveys. The higher figures were from a study location close to Dhaka where both the price of wood and of labour would be affected by higher urban costs. These results, from early 1978 and early 1981 respectively, have to be compared with this project's survey results, from late 1978 (Comilla) and early 1979 (Tangail) given in Table 6.3.

Table 6.3
Average Costs of a Dheki

District	Sample size	% owners giving cost	% of cost figures estimated	Av. cost (Tk)	Av. life (years)
Comilla	996	93	51	66	21.5
Tangail	805	94	41	42	14.5
Overall	1801	93.5	46	55	18.4

Note: If respondents did not have specific knowledge of the current price they were asked to provide an estimated price. 6-7 per cent were unable to provide any answer.
Source: IDS Postharvest Project Survey.

The difference between the two regions can be largely explained by the fact that the Tangail sample came from Madhupur _thana_ which has a large (but shrinking) forest and the trees are suitable species to make _dhekis_ with. The difference between these average results and the other field surveys is more difficult to reconcile but two factors may help to explain them. First, the large samples reported in Table 6.3 included five out of eight villages (but about 85 per cent of households) in the interior where alternative markets for carpenters and for wood were probably more restricted than villages on main roads; second, the other field surveys were concerned specifically with _dhekis_ used by rice processing businesses - _dhekis_ do vary in size and these businesses tend to have the largest ones.

The wide range in field-based cost estimates (Tk 42 to Tk 209), which are not explained by inflation, has a very limited effect on the accuracy of analysis however. Using a twenty per cent interest rate on capital and accepting the 15 year life associated with both these two extreme estimates, the annual capital costs divided by average family annual throughput (see section 6.8) yields a per maund cost of between Tk 0.23 and Tk 1.16; assuming the average of these is a reliable working estimate the investment cost per maund for a _dheki_ is less than Tk 0.70 - at farm level the additional costs for a shelter only negligibly increase this figure. As described below, in comparison to the labour component in _dheki_ operations the investment costs are insignificant and can be ignored (as for example in Ahmed 1982). Moreover, it must be borne in mind that the situation most commonly faced by farm decision-makers is one where they have a _dheki_ already installed and the choice is whether to continue using it or go to the mill; (we have no evidence on the second-hand market for _dhekis_ but the potential second-hand sale value is probably not a significant decision criterion).15/

Turning to the labour costs associated with _dheki_ work, the two figures required are the labour hours per unit throughput and the value of labour per hour. Ahmed (1982, p.115) reports that four earlier studies used a value of 1 woman-day per maund and contrasts it to his own much lower figure of 0.68 woman-days per maund; the difference is attributed to the earlier studies' probable inclusion of pre-milling operations in their labour time estimates for milling. In fact, the difference between his and the earlier results is much greater than he recognizes, since they were measuring time per _dheki_ not time per person; with two to three women working on the _dheki_ the earlier studies actually use a value of at least two women-days per maund. One of these studies (Harriss 1979) lists two further references in support of this sort of value which is approximately 5 lbs (2.3 kg.) per woman hour. Studies of hand-pounding from two other countries (Indonesia, reported in Collier 1979 and Sierra Leone, reported in Spencer et al 1976) give much higher figures, respectively of 5 kg and 4.6 kg. The Sierra Leone study was a carefully conducted time and motion study with a sample size of seventy.

Varietal differences, grain quality or milling level are not sufficiently important influences to explain the time differences and the lower estimates for Bangladesh could be due to a lower efficiency in the dheki compared to hand-pounding or to inherent inaccuracy, e.g. regarding length of a working day, using interviews rather than actual measures.

Therefore, rather than rely on our own questionnaire results, which had given a figure of 8.3 woman hours per maund (n = 538) for this essential figure, a number of physical measurements were taken. After two years residence in our survey villages, when their presence no longer caused a stir, our female investigators sat and timed how long it took to process rice on the dheki. Each worker (up to four were employed) was individually timed and any period not actually spent working was excluded. There were 14 results obtained from Comilla (8) and Tangail (6), and in both districts a mix of family and hired female labour was used. There were no differences between the regions in milling yield which averaged 71.6 per cent (standard deviation 0.03). There were regional differences in the average time taken however. In Comilla the average time per maund was 9.5 hours, compared to 12.2 hours in Tangail. There was no control over variety but this is unlikely to have affected the results; the most likely explanation is of scale economies in labour use (the marginal product of additional labour being greater than the average product). In Comilla, three workers were used except for one case of two and one of four workers whereas in Tangail, except for one case of one labourer, two labourers were used. The average time taken across all 14 results was 10.4 woman hours, and, for convenience, we round this to 10 hours. With an adult male consumption of (say) 250 gms of rice per meal this involves nearly six minutes work per meal; with a family of four adult male equivalents and two meals a day the daily labour requirement for dehusking, polishing and cleaning rice with the dheki is more than 45 minutes. The costs to the farm household of these ten hours of female labour per maund depend most upon whether family women or hired women do the *dheki* work. We have seen earlier that about 17 per cent of total *dheki* labour is hired and for this 17 per cent a reasonably precise labour cost can be estimated. In the case of family female labour however an imputed value has to be derived. This issue is explored at some length in Section 6.10; here we consider the costs to the farmer of the alternative to the *dheki*, the huller mill, to show what the imputed values of female family labour would have to be in order to make the *dheki* the more cost-effective technique.

The farmer can save this ten hours of *dheki* labour by using the rice mill. He incurs additional costs - in the form of his own labour time or for hired labour - on the transport of the grain to and from the mill and for the milling charge. For each of eleven mills surveyed in detail, all customers on two days were asked how far they had come and what transport they used. The vast majority (about 90 per cent) carried their grain in bags on their head, the remainder used country boats, bullock-carts or rickshaws. The one-way distance per customer per day varied between a quarter of a mile and over three and a half miles, and the average distance, weighted by quantity, was approximately one mile. The quantity transported on one head load varied between 20 and 93 kilograms and averaged 56.5 kilograms equal to 1.5 maunds. The average load is 4.5 maund miles per hour at three miles an hour carrying speed and, for costing purposes, the cost of a return journey of one mile with one maund (2 maund miles) is valued at 26.7, say 30 minutes. With milling time and average waiting time the total labour time is much more, probably between three quarters of an hour and one and a half hours - average waiting time is unknown but on occasion was observed to be over two hours. We have adopted a value for total male labour in transport of one hour.

The milling charge in these eleven mills varied between two and five takas per maund at the beginning of the survey (June 1979) and two and six takas per maund at the end (August 1980). During the period the average change in the rate for electric mills was only 20 per cent compared to 27 per cent for diesel mills and the average final rates were Tk 4.8 per maund (diesel) and Tk 2.7 per maund (electric). To some extent the higher charges at diesel-powered mills are compensated by the lower distances to get to them on average since they are much more widely distributed but in other cases these higher charges reflect an element of monopolistic control over milling services in interior villages. The calculations given below on farmer benefits from the mill are average results and farmers further (than average) from the mill or facing monopolistic control over milling would find the dheki relatively more attractive than these average figures.

Using the average agricultural wage rate (without meals) of Tk 12.4 for 1980 to value one hour of labour time and taking the higher charge for diesel-powered mills, the approximate cost to the farmer using the rice mill is Tk 6.4 per maund. Therefore, even after allowing for the half an hour's labour required (see Section 6.4) for cleaning the milled rice and separating husk and bran, the imputed value of family female labour per hour would have to be below Tk 0.68 (about 5 US cents in 1980 prices) for the farm household not to benefit by using the mill instead of the dheki. If, as with male labour, we use the market wage rate to impute the value of female labour it comes to approximately Tk 9 per maund, Tk 0.9 per hour.16/ Given the uncertainties and variations (in distances, quantities, milling rates and wages) these figures are indicative only but their message is fairly clear. At reasonable imputed values for family female labour, and whenever hired female labour is used, the dheki is more expensive than the huller mill.

Once mechanical rice milling becomes available the only constraints faced in using it are cash constraints and male labour constraints. For poor households with underutilised female labour - but labour which would not seek wage employment - it may very well be decided to save the cash expenditure associated with using the mill; the decision is likely to be made in most cases by a male head of house whose labour will not be used in the dheki work anyway. If adult males or strong enough children (of either sex) are not available to carry the grain to and from the mill and the household cannot afford to hire-in labour, then this may prevent rice mill use because women are bound by religious and cultural conventions that awards status if they refrain from appearing in public - and, indeed, refrain from physical labour anywhere. Whilst this does not completely prevent women from taking paddy to local mills - and is less likely to prevent poor women from doing so (see Lipton 1983b, pp.26-33 and Chapter Seven below) they almost always send someone else instead and prefer to do so. It is likely, that for some of the poorest households, labour constraints (male labour availability and opportunity cost) will always be binding and, in other households, that they will be binding seasonally, but the benefits from using the huller mill are so substantial that all households will do so when free from such constraints. Households that hire in female labour for dheki work, especially those with surplus child or male labour, have most to gain, at least in terms of physical product, from rice mill use and, as the short survey reported above showed, (Section 6.2), they are early adopters.

6.6 THE PRIVATE PROFITABILITY OF OWNING HULLER MILLS

Two studies (Harriss 1979 and Jabber 1982) have demonstrated the high profit levels associated with investment in huller mills in Bangladesh. These results were confirmed in a study undertaken in this project, reported in Shahnoor (1980) and summarised here; the relatively longer period in the field perhaps allowed greater accuracy and greater confidence in results but in essence they support the conclusion of the other studies.

An initial survey17/ of thirty-five mills in the two project field areas (Chandina thana, Comilla district and Madhupur thana, Tangail district) was used to obtain background data on milling practices and to enable selection of a sample of eleven mills for more detailed study of two sorts. First, these mills were visited weekly - for over a year in the six Comilla cases and for about six months in the five Tangail cases - to collect information by day on quantities processed, labour costs, maintenance, repair and non-labour operating costs. Most of the mill owners, who were purposively selected on the basis of their cooperativeness as well as their representativeness, kept written records for the study and provided additional information on the reasons for non-operation of the mill.

Secondly, each mill was observed all day on two days (a market and a non-market day) to obtain information on average quantities and average distances for paddy coming to the mill - already reported above - and to check the millers' estimates of quantities processed, mill charges and associated fuel costs.

Table 6.4 gives a summary for the eleven mills of the raw data on quantities processed, gross income, operating costs, capacity utilisation and period of data collection (which should be carefully noted in making comparisons). Table 6.5 gives gross income and operating costs, on an annualised basis, fixed investment costs and rate of return (annual profit as a percentage of fixed investment).

Four mills are shown not to be making a profit and this requires a brief explanation. In the first case, (mill no. 2), the mill suffered a major breakdown18/ which kept the mill closed for five and a half months. In the next two

Table 6.4
Operational Performance of Eleven Bangladesh Rice Mills (1979-80)[1]

Mill[2] No	Quantity Processed[3] Paddy (Tons)	Wheat (Tons)	Gross Income (Taka)	Operating Costs (Taka)	Nos Months Data	Average capacity utilisation (%)[4]
1	58.3	67.6	28,961	23,962	12.9	13.95
2	125.2	24.6	17,429	23,247	13.5	12.24
3	95.4	85.9	39,586	21,480	12.8	35.69
4	214.9	48.4	30,536	25,476	13.5	22.23
5	195.4	149.1	36,214	27,843	12.8	35.92
6	170.8	72.6	22,322	16,240	13.3	16.56
7	6.8	62.1	10,389	14,311	5.7	17.37
8	38.9	46.3	15,257	11,258	6.1	53.00
9	43.8	20.0	5,347	8,019	5.7	10.59
10	10.6	37.0	11,064	11,800	5.5	22.69
11	28.5	59.5	9,504	7,373	5.7	33.54

Notes: 1. Mills 1-6 are located in Chandina Thana, Comilla district and 7-11 in Madhupur Thana, Tangail District.
2. Mill Nos. 5,6,7,9 and 11 were electrically powered for part (7 and 9) or all (5,6 and 11) of the data collection period. The remaining mills were powered by diesel engines.
3. All data was collected through weekly visits in which the previous seven days operations were recorded.
4. Actual hours of utilisation were calculated by dividing the total quantities of paddy and wheat processed by the average number of maunds per hour, for paddy and wheat respectively.The quantities were obtained from the weekly survey and the hourly throughputs were based on physical measurements taken at each mill. The percentage share of paddy and wheat in actual hours was used to estimate their share in potential hours (total 3,600). Multiplying by the respective per hour throughput figures and summing the two results gave the cereal throughput potential, allowing calculation of capacity utilisation.

Source: IDS Project Survey.

Table 6.5

Annual[1] Economic Performance of Eleven Bangladesh Rice Mills

Mill No	Annual Gross Income	Annual Operating Costs	Gross Profits	Fixed Investment Cost[2]	Rate of Surplus (Percentage Return on Fixed Investment)
1	26,940	22,290	4650	17,900	25.98
2	15,492	20,664	-5172	9,776	-
3	37,112	20,138	16,974	20,815	81.55
4	27,143	22,645	4498	19,650	22.89
5	33,951	26,103	7848	23,570	33.30
6	20,140	14,653	5487	24,300	22.58
7	21,871	30,128	-8257	28,350	-
8	30,014	22,147	7867	22,300	35.28
9	11,257	16,882	-5625	17,700	-
10	24,140	25,745	-1605	22,000	-
11	20,008	15,522	4486	22,600	19.85
Average	24,370	21,538	2,832	20,815	34.5 (10.9)[3]

Standard Deviation	7,350	4,497	7,036	4,520	19.9 (36.3)

Notes: 1. Annualisation of the Madhupur Mills (7-11) data from the six months or so raw data ignores seasonal influences; these may mean that the data collection period (March-August) is untypical of a full year's performance. Indeed, the relative share of wheat is almost certainly overstated since the wheat harvest starts in March. However, there did not appear to be any basis for readjustment which would clearly and significantly improve the representativeness of these mills data by adjusting proportionate shares. As the text discussion suggests though, there is good reason for thinking that the underlying profitability is under-represented in at least two of the three loss-makers amongst the Madhupur Mills.

2. These investments are unadjusted for inflation or shifts in relative prices even though the mills were actually set up at different times over the period 1967 to 1979. As discussed in the text the average return for the six Comilla mills (nos. 1-6) at replacement cost for the fixed investment, is 19 per cent.

3. The figures in brackets include the four loss-making concerns which as the text describes, were mainly 'abnormal' cases, eg because of lost operating time during a change of power source from diesel to electricity.

Source: IDS Project Survey.

cases (7 and 9) the simple annualisation from 6 months data is misleading since both these mills suffered a closedown of operations whilst converting from diesel to electrical power during the survey period. The final loss-making mill is a road-side located diesel mill which for the first time was facing competition from electrically- powered mills located nearby and offering cheaper rates (Tk 2/3.5 per maund compared with Tk 4, for paddy and Tk 5/8 per maund compared with Tk 10, for wheat); the owner has several business interests which, although we have no data only impressions, may have led to poorer management of the mill. He plans to electrify as soon as it becomes possible at his particular location.

The remaining mills enjoyed an average rate of return of nearly 35 per cent (n=7) - this result is strongly influenced by one mill with a return of over 80 per cent;19/ the average return for the other six was nearly 27 per cent which is still very respectable. These results are based on historic rather than replacement cost for the fixed investment. For the six Comilla mills replacement cost has been calculated (for the SBCA given in the following Sections) at Tk 29,917; using this value for fixed investment, these six mills have an average return of just over 19 per cent.

It is evident from the results that in a situation of satisfactory average rates of return there are substantial variations. We have noted the abnormal conditions for the loss-making concerns that resulted in low capacity utilisation; the remaining mills achieved their profits despite their also having very low capacity utilisation - exceeding forty per cent in only one case. Using annualised values, a total for the eleven mills of 763 days were lost and the causes were mechanical problems (36.7%), lack of spare parts (26.3%), electricity failure (18.4% but 63.9% of days lost by electrically powered mills), lack of grain supply (14.8%) and other reasons (3.8%). These figures only show whole days lost but on many other 'working' days throughput was small due to lack of supply and temporary breakdown or power failure. These figures suggest that whilst the effective capacity is much lower than installed capacity there is still substantial surplus capacity and profitability will depend upon considerations such as the paddy-wheat mix, locational and regional differences and product quality. From this small data set, there is little additional useful knowledge that can be directly derived to indicate the impact of these considerations on private profitability; moreover, it is likely that fine-tuning of results, as a consequence of incorporating in more detail these, or other considerations, would not alter the general direction or strength of the conclusion that under a wide range of operational efficiencies, the huller mill is an attractive investment for rural entrepreneurs. However, before turning to the analysis at accounting prices it is necessary to comment briefly on the main change occurring in rice milling technology, the shift from diesel to electric power.

Table 6.6
Performance Differences - Diesel & Electric Rice Mills

	Average of Four Diesel Mills	Average of Three Electric Mills
Variable Costs per Annum (Tk)	20,543	17,182
% Share of Fuel in variable costs	48.2	26.2
Throughput per hour: Paddy (tons)	0.35	0.60
Wheat	0.11	0.16
Throughput per annum: Paddy (tons)	103	132
Wheat	69	110
Cost per ton of throughput (Tk)	119	71
Av. Rated Horse Power	11.25	25
Av. capacity utilisation %	31.2	28.7

Source: IDS Project Survey.

Electric mills have both lower investment costs20/ and lower operating costs (see Table 6.6) per unit of throughput. Wherever electricity becomes available, diesel mills will inevitably be displaced. However, both technologies will remain important since large parts of the country will remain non-electrified for the immediate future.21/ Even though electric mills, per unit capacity, may have a larger catchment area because they pass on to customers some of their cost savings - thus reducing customer cash costs and leaving more room for transport costs - they are likely to be more closely clustered;

overall, therefore, it is not clear whether electric mills will operate at a higher capacity utilisation, and, if they do, whether this would result in a more rapid rate of replacement than one electric mill for one diesel mill. This depends on the intensity of clustering and upon whether the increase is principally due to greater density of customers within the same radius or an extension of the radius of the catchment area. The relative profitability of diesel and electric mills is also affected by these considerations.22/

The detailed field data for the seven mills in the sample operating under normal conditions allow a comparison of both capacity utilisation and profit levels; these results must be regarded as illustrative rather than representative given the small sample size. The four diesel mills averaged a 41.4 per cent return on capital with average annual profits of Tk 8497 and 31.2 per cent capacity utilisation. The three electric mills averaged a 25.2 per cent return on capital with average annual profit of Tk 5940 and 28.7 per cent capacity utilisation. In this sample it would appear quite plausible that for electric mills greater competition has reduced both capacity utilisation and profits, though profits remain reasonably high. Diesel mills, now only competitive in the rural interior, are continuing to earn high profits by virtue of their relative isolation and consequent ability to charge higher rates. Overall, it would appear that customers will benefit but no fundamental shifts in the profitability of the sector will result directly from the process of electrification. These conclusions are based on the survey results using market prices; using social accounting prices, the incentives for use of electrified mills are probably somewhat diminished because of a relatively greater subsidisation of electricity which enhances the profitability, at market prices, of electrically-powered mills.23/

6.7 SOCIAL BENEFIT COST ANALYSIS: AN OUTLINE OF THE APPROACH

Chapter Two outlined the objectives and methods of Social Benefit Cost Analysis in a postharvest context. Merit wants relevant to Bangladesh - employment and food self-sufficiency - were briefly discussed in relation to SBCA and the principal purpose of SBCA was identified as the maximisation of aggregate consumption, taking account of distribution amongst present consumers and between them and future consumers. In Chapter Five some boundary values were offered for critical variables in the threshing choice of techniques to show that private preference for pedal threshers probably did not conflict with social preference measured through SBCA. So far, in this chapter, it has been shown that the huller mill is a very profitable private investment yet it is quite common for the disadvantages to female wage labour to be given as a reason for discouraging the influx of hullers. The substantive question at issue in this section is whether social profitability, measured through an SBCA application to the comparison, would reverse the choice of technique. The methodological question is whether it is possible to estimate social values for the inputs (labour, fuel, mills, dhekis etc) and for the output (the milling service) that provide a more accurate reflection of the aggregate consumption value than do market values.

At first sight it would appear probable that SBCA must necessarily reverse the choice; if, as is the case in Bangladesh, rural female workers are underemployed and come from poor homes, then the shadow wage rate is low and, quite possibly, could be taken as zero. Since there are no other operating costs except this labour, and since _dhekis_ are cheaper than mills, the _dheki_ must be socially more valuable. Whilst this argument has a lot to commend it, not least its clarity, there is one compelling reason, and several other supporting reasons, why more judicious analysis is required. 1. The reversal (in cost effectiveness rankings) is based upon the zero social opportunity cost of female labour; whilst this almost certainly holds for _hired_ female labour it may not be true for _family_ female workers. It only requires a small shadow wage to affect results since the _dheki_ uses sixty times24/ more labour than the mill and over 80 per cent of this is family labour. It must be admitted that none of the reasons, given in Chapter Five, for placing a relatively high value on male family labour during threshing apply particularly to the _dheki_ case which, for a lot of the time, uses off-season female labour. The value is nevertheless thought to be positive and an estimate is presented below. If it is positive a more careful SBCA is relevant because of two further adjustments to the private profitability analysis which involve substantial departures from market values. These are: 2. the use of an accounting price for private investment; and 3. use of different consumption weights for different income groups.

Also, since the change in technique is occurring fairly rapidly a real reversal would involve substantial changes in government policy and patterns of intervention that are themselves not without cost - these too must be taken into account. To put it another way, the only losers as a consequence of the introduction of hullers are hired female labourers - should the government intervene and if so, how? (Details on the earnings structure of female wage labour is given later on (Chapter Seven) and for the moment it is simply accepted that, with an income per consumption unit only slightly over Tk 1000 per year, these 'losers' come from some of the poorest households in rural Bangladesh and that the output foregone by using their labour is zero). The substantive conclusions we came to, some pages hence, are concerned very directly with the view that socially, the cheapest policy for the government is to intervene through programmes _for_ poor women rather than _against_ huller mills. As Sen's (1984, pp.224-241) discussion illustrates the conventional practice of SBCA may easily fail, in the assumptions used about accounting prices especially, to take account of the control areas actually accessible to planners.

One normal approach to SBCA in this context would be to regard the huller mill investment decision as a 'project' with no incremental output but as a saver of resources. The present social value of the investment would be calculated as the net benefits25/ going to the different income groups - including the savings on _dheki_ investment no longer needed - less the costs associated with consumption foregone by the losers. The 'project' would be worthwhile if the present social value was positive, i.e. it is a net saver of resources. An alternative approach, analytically identical but practically and presentationally simpler, is adopted here by regarding the two techniques as alternative 'projects', and choosing the one with the higher PSV.

Ideally, the derivation of accounting prices in a specific analysis should be consistent with those used in other planning activities. The most common use and the most accessible source of accounting prices for Bangladesh is in World Bank project evaluations and 13 sets of project accounting prices, from 13 different projects, were available to the author.26/ Various references will be made to these data but it has been necessary to develop some accounting prices, notably for the major inputs, private investment funds and labour, since relevant figures were not available. A World Bank Standard Conversion Factor (SCF) of 0.75 has been adopted, to estimate accounting prices for spares and repairs and parts of other items; the Accounting Rate of Interest (ARI) used (12 per cent) is also a World Bank figure but the comparison is fairly insensitive to the ARI since the investments are assumed to have constant annual benefits and equal lives.27/ Since values established and refined through usage for the major inputs and, critically, for consumption weights do not exist and since their derivation involves value judgements, as well as inference from data, it would be presumptuous to suggest one set was unambiguously correct. To deal with this problem, switching28/ values of critical parameters are identified and assessed in 6.11.

The analysis, following convention, proceeds in three steps: i) A comparison of the techniques at market prices. ii) A comparison at efficiency prices which measures inputs and outputs at their social opportunity cost. iii) Adding distributional considerations into the analysis by deriving a comprehensive estimate of the relative social worth of the income streams resulting from the investment. (The generic term, accounting prices, refers to either efficiency or social prices). The point of this division is that it allows separation of adjustments to market prices according to efficiency criteria and equity criteria.

Efficiency prices correct for differences between market prices and their social opportunity cost - but all incomes are regarded as equally valuable whether they enhance public or private consumption or savings. Social prices offer a more precise specification by taking into account the relative weight given to consumption and savings - in effect a valuation of a stream of future consumption benefits in terms of present consumption - and taking account also of the different weight attached to the marginal consumption of different income groups.

There is no shortage of texts and case studies on SBCA29/ though, to the author's knowledge, there is no comprehensive application to the choice of techniques here in question. The methods followed here are based, most immediately, upon Little and Mirrlees (1974) - though occasionally employing terms used by other manuals but still based on Little-Mirrlees methods - because their approach explicitly and actively incorporates issues concerning the relative value of private investment and savings which is a major topic in this comparison and which other manuals give less emphasis to.

The main departure from Little and Mirrlees (1974) is in the estimation of b, the base consumption level at which additional income is regarded as equally valuable as uncommitted social income; hopefully, the departure which is described below represents an alternative for Bangladesh which allows more useful analysis. Throughout, the emphasis is upon practical application rather than theoretical exposition of SBCA methods.

However, it is as well to recognise from the outset that an absolutely comprehensive quantification of the aggregate consumption objective, the main thrust of SBCA, is not possible. One reason for this concerns the constraints on effective intervention which, as discussed briefly above, is an issue which is largely by-passed by SBCA. A second, and in our context, very important reason is the assumption underlying SBCA of a well-functioning labour market in which labour displaced can seek new sources of earnings as effectively as any other labour. This is not true for landless wage labour women in Bangladesh where loss of dheki work represents a serious threat to their whole livelihood. In the light of the quantitative results this topic is explored more substantively in Chapter Seven.

6.8 A COMPARISON AT MARKET PRICES

The rice mill survey data shown earlier established that for a range of capacity utilisations the huller mill is privately profitable.

The estimates of social worth presented below are based on the results for the six of these rice mills from Comilla, for which there is a full years data from the weekly survey. The Comilla sample set provides appropriate30/ representation of the electric-diesel mix with four diesel mills and two electric; they reflect the full sample range of capacity utilisation and profitability. (See data for mills one to six in Tables 6.4 and 6.5). The data set available could anyway provide little insight into possible regional characteristics of the social and economic impact of huller mills. (Some insights can be derived from the data on the income and work of female wage labour which are presented in Chapter Seven). For policy purposes, it seems likely that the timing rather than the character of mill diffusion is the main regional influence - due to differences in infrastructure, volume and proportion of marketed surplus and land size distribution. To the extent that these differences matter and can be separately analysed it is an advantage to have the six mills from one district. The main reason for this selection however is the fact that the detailed data on operational performance from the weekly survey is available for a full year of operations so that there is no ambiguity concerning the interpretation of the baseline information. Since the analysis involves estimation of values for several other key parameters it is particularly important to do whatever possible to maintain a high degree of confidence and interpretability for those parameters where solid information does exist.

The average values for fixed investment and operational performance are given in Table 6.7, based on the individual mill data from the survey in Comilla. These values, with an adjustment for the paddy-wheat mix as described below, are compared to dhekis being used for a similar volume of throughput.

Table 6.7
The Comilla Mills: Investment and Annual Operational Performance Averaged At Market Prices

Fixed Investment			19,335 Tk
Annual Gross Income			26,796 Tk
Operating Costs:			
	Fuel	8688	
	Repairs and spares	8413	
	Labour	3981	
	Total		21,082 Tk
Value added (net of labour)			5,714 Tk
Annual Rate of Return			30 per cent
Annual Quantity Processed:			
	Wheat	70	
	Paddy	130	
	Total		200 tons
Capacity Utilisation a/ 23 per cent			

Note: a/ Based on 12 hours operation for 300 days; see Table 6.4.

Source: IDS Project Survey.

The fact that virtually all huller mills also share their power source with wheat grinding stones (chakki) means that the social valuation of rice milling, as in private profitability calculations, cannot be divorced from wheat grinding. It is necessary to accept therefore that grinding wheat and milling rice is a joint investment decision. The fact that all the sample mills conformed to this pattern and that, with few exceptions, the same situation prevails elsewhere in the country for these small mills is reasonable validation.31/

The total throughput of 200 tons includes 130 tons of paddy. It is not correct though to ascribe 65 per cent of the investment and operating costs to paddy. The explicit approach adopted here is to presume that the share of wheat and paddy in terms of fixed investment, operational cost and profitability is accurately reflected in their shares of operating time. The implicit presumption is that the alternative to wheat grinding at huller mill sites bears the same social value profile as the dheki does to the huller. This is almost certainly false and the direct opposite is probably true; i.e. the alternative to wheat grinding at huller mill sites is a more capital intensive operation with a markedly different social value profile compared to the dheki. Nevertheless, this approach remains useful in isolating the direct impact of the hullers. Operating hours for wheat and rice were estimated as the quantity of each cereal processed during the year divided by the respective average measured throughput per hour. These average throughputs for the six Comilla mills were 3.87 maunds an hour for wheat and 11.78 maunds per hour for paddy.32/ Therefore paddy operations were accountable for 38 per cent of the annual income and operating costs in Table 6.7.

The same basic approach can be applied to the fixed investment costs, with two adjustments. The first adjustment is to exclude items specific to wheat grinding and to include the full value of those items specific to paddy hulling (i.e. the huller itself). These sets of items are in fact fairly similar in price and anyway only a small proportion of total investment costs so this adjustment has been ignored. The second adjustment is to take cognisance of the fact that the investments in these six mills did not take place in 1980, when operating costs and income were estimated, but at different times during the period 1972-1979. Presuming there have been no changes in relative real prices, the wholesale price index for industrial products provides a reasonable basis for re-estimating these investment costs at 1980 prices. The unadjusted share (0.38) of paddy in total fixed investment (Tk 19,935) is Tk 7347; adjusting each of the six mill investments by the relevant price indices (GOPRB 1983a, p.355) to 1980 prices gives an average figure of Tk 11095 33/ which is the market price of fixed investment that has been used. This is equivalent to a total investment cost at 1980 prices of Tk 29,197. Based on Table 6.7 and these assumptions regarding the share of paddy in total milling, Table 6.8 gives the market price values at 1980 prices, used in the basic calculation.34/

In addition to the mill operating costs there are two family labour inputs associated with mill use; male labour in transport to and from the mill and female labour in recovering rice from the by-products and in separating the husk and bran. These tasks were shown to require, on average, an hour of male labour and half an hour of female labour and these inputs have to be accounted for in assessing the annual value added from adoption of the mill. At this stage all labour inputs are valued, notionally, at market wage rates. The 1980 rate for male labour was Tk 12.4 per day (GOPRB 1983a p.364) and, assuming an eight hour day, TK 1.55 per hour; with a total of 130 tons (3482.7 maunds) milled the labour cost of transporting the paddy is Tk 5398. The female wage labour rate (see footnote 16) was Tk 0.9 per hour; the cost of female labour for their post-milling work was therefore Tk 1567.

Similar figures for an equivalent throughput using the _dheki_ are also given in Table 6.8. The _dheki_ fixed investment costs are based on the number of _dhekis_ required to process 130 tons of paddy per annum and the average investment cost per _dheki_. This average cost has been estimated as the mid-point of the range of estimates from field surveys on _dheki_ costs and equals Tk 126 (see Section 6.5). The number of _dhekis_ depends upon the average annual throughput which has been calculated as follows. Average per capita cereal intake (adjusted for age, sex and activity group) has been estimated as 523 grams per day of which 94 per cent is rice (Ahmad 1977). The _dheki_ milling ratio is 72 per cent which yields an average requirement of 6.7 maunds of paddy per person per year. With an average family size from the Comilla household survey of 5.65 the household annual throughput is 37.8 maunds; making a small allowance (2.3 per cent) for special occasions when consumption may increase this yields a throughput of 38.67 maunds or 1.44 tons per _dheki_ per year; this figure has to be adjusted to make allowance for _dheki_ use for the consumption of households that don't own their own _dheki_;35/ this results in an increase of 0.37 tons per _dheki_ giving an average throughput of 1.81 tons. For 130 tons of paddy 72 _dhekis_ will be used and at the average cost of Tk 126 will result in a fixed investment cost of Tk 9072. This approach takes explicit recognition of the fact that practically all the marketed surplus is already mechanically hulled and the policy issue concerns the use of rice mills for rice consumed by family-farms.36/

Table 6.8
The Choice of Technique: A Comparison at Market Prices

		(Takas)
	Dheki	Huller Mill
Nos. of Units	72	1
Annual Throughput (tons)	130	130
Value of Output (taka)	10,182	10,182
Fixed Investment	9,072	11,095
Annual Operating Costs (taka)		
Female Family Labour	26,015	1,567
Male Family Labour		
(i) Mill Owner		1,513
(ii) Farmer		5,398
Female Hired Labour	5,329	0
Male Hired Labour	-	1,513
Fuel	-	3,301
Spares and Repairs	-	3,197
Annual Operating Surplus	-21,162	-6,307

Note: The costs of family labour are notional and at the next stage of analysis, at efficiency prices, they are substantially revised. Here, family labour has simply been valued at the wage labour rate.

Source: Table 6.7 and text.

The dheki operating costs consist entirely of labour37/ at the rate of ten hours per maund. 130 tons is 3482.7 maunds and following the analysis of Section 6.2 it is assumed that 17 per cent (=5921 hours) of labour is hired and the balance of labour requirements (28906 hours) is provided by family female labour. The (purely notional) wage rate used to price female labour at this stage is taka nine per maund (see Section 6.5). All these numbers are summarily listed in Table 6.8. [For ease of comparison, Tables 6.8, 6.9 and 6.10, giving the choice of techniques comparison at market, efficiency and social prices respectively, have been placed together; the derivation of Tables 6.9 and 6.10 is given in the next two sections.]

Table 6.9
The Choice of Technique: A Comparison at Efficiency Prices

(Takas)

	Dheki	Huller Mill
Value of Annual Output	10,997	10,997
Fixed Investment	7,076	7,787
Annual Operating Costs		
a. Female Family Labour	9,211	261
b. Male Family (mill owner) Labour	-	1,149
c. Female Hired Labour	1,007	-
d. Male Hired Labour	-	1,149
e. Fuel	-	3,235
f. Spares and Repairs	-	2,398
Annual Operating Surplus	779	2,805
Net Present Value a/	-1,770	11,318
Benefit - Cost Ratio b/	0.75	2.45

Notes: a/ Net present value is the product of the annual operating surplus and the present worth of the annuity factor less the fixed investment costs. The annuity factor in both cases is 6.811 based on an ARI of 12% and 15 years life.
b/ The Benefit-Cost Ratio is the product of the annual operating surplus and the annuity factor divided by the fixed investment costs.

Source: Table 6.8 and text.

Table 6.8 shows that, when all the different inputs associated with each of the techniques are valued at market prices, neither technique has a positive gross operating surplus. This feature deserves some comment since it seems to suggest that neither technique is viable and yet both techniques are in wide use. There are two explanations. First, the market prices used to impute values of inputs, especially labour, may overestimate their opportunity cost; most of the analysis in the next section is concerned with this question.

Table 6.10
The Choice of Technique: A Comparison at Social Prices

(Takas)

	Dheki	Huller Mill
Value of Annual Output	10,997	10,997
Fixed Investment	5,307	3,193
Annual Operating Costs		
a. Female Family Labour	9,211	261
b. Male Family (mill owner) Labour	-	1,497
c. Female Hired Labour	-10,056	-
d. Male Hired Labour	-	868
e. Fuel	-	3,235
f. Spares and Repairs	-	2,398
Annual gross operating surplus	11,842	2,738
Net Present Value a/	75,349	15,456
Benefit - Cost Ratio b/	15	6

Notes: a/ Net present value is the product of the annual gross operating surplus and the present worth of the annuity factor less the fixed investment cost. The annuity factor in both cases is 6.811 based on an ARI of 12% and 15 years life.
b/ The Benefit-Cost Ratio is the product of the annual gross operating surplus and the annuity factor divided by the fixed investment costs.

Source: Tables 6.8 and 6.9 and text.

Secondly, the value of output may be underestimated. The output, milling services, has been valued at the cost charged by millers to farmers which must be an underestimate of the real value. Since farmers are prepared to pay the milling charge and provide the labour to clean the rice afterwards they clearly place a value on milling greater than the collective cost of all these inputs. Sen (1984, p.412) quotes Scitovsky on this issue: "The consumer gets more satisfaction out of anything he buys than out of the money he pays for it, otherwise he would not buy it". The same reasoning applies of course to labour use. The degree to which the consumer benefits more than the price he pays is the extent of consumer surplus, which is unknown but it must be positive. However, accepting that prices paid 'reveal preferences' market prices are nevertheless accurate indications of relative values.

The problem in this case is that the only alternative basis to valuing the milling service, in the absence of a comprehensive market price, would simply be to add up all the input costs which of course would leave the output value exactly equal to the costs of obtaining it. (Arguably, in this case, the paddy-rice price difference could be used to value output but this price difference reflects other cost factors (transport, storage, insurance and risk) as well, depending on where the figures are obtained from, and offers no very great precision for the valuation of milling for farm-level consumption paddy.)

In our context, it does not really make any difference what estimated market value of output is since both techniques produce the same physical output; both techniques show positive operating surpluses at efficiency and social prices and there seems to be no good reason for trying to derive the (true) higher value of milling services at market prices when the results are unaffected.

6.9 A COMPARISON AT EFFICIENCY PRICES

Output Curiously, the only item for which the two techniques share a common accounting price is the value of unit output. At accounting prices they share no identical inputs (other then the half hour of female family labour cleaning the milled rice) since they draw their capital, material inputs and labour from different sources. The mill is owned by a rural entrepreneur using fuel, male labour and factory produced material inputs; the _dheki_ is owned by a family farm, operated by women and made by village carpenters. And, since the two techniques are being compared at the same level of throughput, the output price is of limited relevance. However, in Section 6.4 it was suggested that one of the values to be incorporated in the sensitivity analysis was output since the value of huller output may be slightly lower than that of the mill. So, for the sake of completeness, and to allow a statement of benefit-cost ratios, an efficiency price for output has been estimated. Milling services are non-traded and marginal social benefit is equal to their price, since this is presumed to reflect consumers willingness to pay, multiplied by the consumption conversion factor which (see below) is estimated to be 1.08 at efficiency prices. Thus, the efficiency price for output is Tk 10,997.

A further assumption, which greatly simplifies presentation, is that the mill, as the dheki does, has a 15 year life - and, therefore has an equal life-time value of output. The huller unit itself has a much shorter life but as the various parts fail they are replaced and all these costs are reflected in the spares and repairs item in operating costs. The foundations etc for securing the engine could normally be expected to have a 15 year life. However, in some project analyses (e.g. the Bangladesh Power Development Board 1977) the engine is assumed to be of shorter life. This is felt by us to be wrong because diesel engines and electric motors are maintained very much like the huller itself with parts being replaced or repaired as required rather than whole new units being purchased. Eight of the eleven mills in the sample were using second-hand engines or motors some of which were over 15 years old. In other words, the spares and repairs item includes substantial maintenance costs that extend the normal life of the power sources. Moreover, the scrap value of engines is high relative to their costs, in one example38/ as high as two-thirds; even discounted at 12 per cent over 15 years this is equivalent to ten per cent of the purchase cost or one and a half years of engine availability and yet the allowance actually made is only 7 per cent and only on machinery costs (see below). For these reasons, a 15 year life is a perfectly plausible approximation.

A final consideration on estimation of the life-time value of output is the possibility that the unit value of output may change over time. For example, it is possible that with the development of the infrastructure, especially roads, provision of modernised transport and electricity, greater competition, eventually followed by rationalisation, will result in higher capacity utilisation and possibly scale economies with larger mill units though this is a distant prospect. Such developments will reduce milling costs. It is unlikely that their influence will be easily discernible as other adjustments occur, including greater price sensitivity to product quality and packaging. However, it does suggest that there could be an increase in consumer surplus which properly should be considered a benefit of the huller. No attempt is made to include this adjustment quantitatively but it should be considered as some (probably fairly small) weight in favour of the huller.

For the huller mill there are four inputs (labour,

fuel, spares and repairs and fixed investment) and for the dheki there are two, female labour and the fixed investment, these are taken in turn.

Huller Mill Inputs

LABOUR: In each of the six Comilla mills there was a labour force of two, (slightly less than the average labour force of 2.4 reported in the larger mill survey - see Section 6.3), consisting of the rural entrepreneur working alongside a hired driver (mill operator). The entrepreneur provides the capital and the management for the mill but he also provides half of the operating labour. The returns to the mill owner are his profits which are a result of all three of his inputs: capital, management and labour. The opportunity cost of the capital input is the social value of the profits he would have earned elsewhere and the treatment of this is dealt with in the next section.

However, mill owners' operational and management labour also has an opportunity cost (even though he does not receive any direct payment for his work). It is not straightforward to calculate the social value of whatever management and labour work the miller may have been doing if he was not busy milling. An alternative approach could be to impute a value for the opportunity cost of his labour and management based on the costs he is saving by not employing a salaried manager. This does not take us much further forward without any evidence (it is a very rare event for these small mills to use a salaried manager and we came across no cases) on what these levels are. A simplifying and, we believe, acceptable alternative is to recognize that his management input should be treated as complementary to his deployment of capital and therefore for his profits to be seen as a return to management and capital; this assumes the entrepreneurial input is committed when the capital is committed. Treating management and capital together is correct provided that the entrepreneur's most profitable alternative uses of capital involve similar complementarity. Likely avenues of investment, eg trade, clearly would involve complementarity between capital and management.

On this assumption (which is the method recommended in UNIDO 1978 p40) the mill owners' management input is rewarded along with his capital input by profits and we are free to value the mill owner's labour input purely as labour so far as its efficiency price is concerned. This is the procedure followed here and the social opportunity cost of hired mill labour is also used to value the labour input of the mill owner/operator.

For hired labour, which has a market cost of Tk 1513, the conversion factor is derived from the ratio of shadow to market wages and the shadow wage is defined as: $SWR = m + (I - \frac{d}{v})(c-m)$.

c = the consumption generated by employment which is measured by the market wage-rate at the consumption accounting price.
d = the value of consumption at consumption level c relative to that at the base level of consumption.
m = the marginal value product of agricultural labour at accounting prices.
v = the premium on savings relative to consumption.39/

The first term on the right-hand side measures output foregone as a consequence of the new job. The second term measures the net costs to aggregate consumption of using resources for current consumption rather than investing them to generate future consumption. This second term is the adjustment from efficiency prices to social prices and is dealt with in the next section. At efficiency prices the shadow wage is simply m, the social opportunity cost of rural male labour.

At market prices, the mill labour wage is Tk 1513 (Table 6.8); as described in the following paragraphs there are three adjustments to this market wage to obtain the efficiency wage, m. The first is to assess the ratio, at market prices, of the marginal product of labour to the market wage; this ratio turns out to be 0.67. The second is to estimate the conversion factor for expressing the market wage at accounting prices; this factor turns out to be 1.08. The net effect (of multiplying by 0.67 and 1.08) gives an efficiency wage of Tk 1095. The third adjustment is to add in the costs of the additional dietary energy intake required by the mill worker in order to undertake the work. This is estimated to be Tk 54 and gives an efficiency wage of Tk 1149 (1095 + 54).

1. It is sometimes assumed for rural Bangladesh that the marginal product of rural labour is zero since there is much more agricultural labour available than can be absorbed by the agricultural sector. This assumption accepts that nothing of social value is foregone as a consequence of agricultural labour devoting time to a new task. However, the seasonal pattern of labour use in agriculture means that

estimation of 'surplus' labour provides a poor guide to the impact on agricultural output of withdrawing labour (Sen 1975, chapter four). It must be recognised also that the poverty of most hired labourers discourages (prevents) long periods of unemployment; indeed, the very poorest rural households are precisely those that cannot, for a variety of reasons (Lipton 1983b, p.15)40/, participate sufficiently effectively in the labour market. It would be inappropriate here to pursue this much-debated issue at length and, in agreement with World Bank studies which all employ a non-zero marginal product for agricultural labour, it is accepted that some output can reasonably be considered to be lost as a consequence of job creation.

In the absence of econometrically-derived estimates of labour's marginal product in agriculture, Little and Mirrlees (1974, pp.277-279) suggest an alternative approach to estimate the social opportunity cost of output foregone. Their suggestion involves two steps: a) obtain the marginal product per day worked either by assuming it is equal to the agricultural wage rate or by assuming it is equal to half the average value added per man-day of employment in agriculture. The Bangladesh agricultural wage rate in 1980 averaged Tk 12.4 (GOPRB 1983a, p.365). Total value added in cereals production in 1980-81 was Tk 60,767 million and the associated employment was 1844 million man-days;41/ according to these figures average value added per day worked was Tk 32.95 and, on Little and Mirlees assumption that the marginal product can be taken as half of the average product, the value of marginal product per day worked is Tk 16.5. Thus,we have two, very crude, estimates of marginal product per day worked and we take the average value, Tk 14.5 to use in the second step. b) The second step is to multiply the marginal product per day worked by the number of days worked in order to derive the annual social marginal product of labour. It has been estimated (IBRD 1983 p.II) that, on average, agricultural labour, in 1980, found employment (both crop and non-crop) for 185 days a year. Using this figure, the annual marginal product of an agricultural worker is 185 X 14.5 = Tk 2683.

There is little which is satisfactory about this method of estimation to derive the social marginal product of labour. The wage rate is highly variable, by crop, region and season, and the labour market is less than perfect with various forms of tied labour which create anomalies in the structure of wages. Also, the assumption that the marginal product is half of the average product has no real validation other than that we know the marginal product is less, probably considerably less, than the average product of agricultural labour.

Ideally, data on time disposition of the labour force would be available and allow accurate estimation of days worked out of days available for work. The manpower survey conducted by the government (the results of which are summarised in IBRD 1983) only uses 'reference weeks' which are difficult to interpret on an annualised basis. IBRD (1983 pp.12-14) reviews several micro-estimates on time disposition but the different assumptions made in the various studies, on labour use in crop production, on participation rates and on the duration of work (hours in the day and days in the year), inhibit consistent conclusions. The value used here, of 185 days for average days worked per participant in the labour force, is IBRD's best estimate after reviewing all the evidence; it is an estimate which, common sense suggests, is the right order of magnitude but there is clearly room for improvement in the data base from which it is derived.

Annual wages for hired mill workers in 1980 were Tk 3981 (Table 6.7) so the implied conversion factor to the social value of marginal product is $\frac{2683}{3981} = 0.67$. Values of 0.50, 0.67 and 0.75 were used in three different IBRD appraisals in 1980 (see footnote 26) so our conversion factor is at least reasonably consistent with other studies. Applying ths conversion factor to the mill workers wages of Tk 1513 from rice processing (Table 6.8) gives a value of Tk 1014.

2. Adjusting this estimate of the social marginal product of mill labour from market prices to accounting prices42/ is done through the consumption conversion factor. It is estimated as follows.

About 75 per cent of total rural incomes is spent on food (IBRD 1980) and about 94 per cent of calorie intake comes from rice (Ahmad, 1977). Expenditure surveys show (IBRD 1980, p.191) that about two-thirds (calories are not everything needed) of total food expenditure and half of total (food and non-food) expenditure is accounted for by rice. The conversion factor for paddy has been estimated (1980) by the IBRD at 1.41.43/ It was observed from the same source that the conversion factors for other crops were much lower, and for consumer items such as chillies and vegetables only 0.75 - in fact it appears in these studies that the Standard Conversion Factor (SCF), also 0.75, has been applied to non-rice food expenditure and also to the quarter of total expenditure which is non-food. Since ultra-precision is not important, and anyway not possible, use of the SCF seems a reasonable procedure; Scott, MacArthur and Newberry (1976) in a seminal study, provide detailed estimates of consumption conversion factors and, having done so, advocate short-cut methods whenever they can be safely employed.

Therefore to adjust market wages to accounting prices the conversion factor is simply the average of 0.75 and 1.41 which is 1.08. And m, the marginal social opportunity cost of labour is equal to the product of Tk 1513 (the market wage), 0.67 (the ratio of market wage to marginal product value) and 1.08 (the consumption conversion factor); this comes to Tk 1095 which is the social value of the output foregone by employing mill labour.

3. Several other considerations can be introduced into the estimation of the efficiency component of shadow wages. Some, such as additional transport costs (especially for urban jobs) or increases in wage rates for other workers, are not relevant in our context. Others, such as a skill premium for mill workers' labour or an allowance for the social cost of the disutility of labour effort, could be considered relevant but are estimated to be of little practical significance.44/ There is one consideration which has not so far been included in the estimate of output foregone and which does need to be included - the cost of providing the additional food consumed as a consequence of energy expended on work. This cost, as will be shown below (pp.331-332), is a significant component of the efficiency wage of dheki labour so it is only correct to provide some estimate of its value in the case of the mill - even though there are far fewer labour hours.

We know (Table 6.7) that capacity utilisaton in the Comilla mills used for this analysis is 23 per cent which is 828 hours a year or 16 hours of work a week. Of this 16 hours, thirty-eight per cent (see p.308) is rice milling (as opposed to wheat grinding) work. Mill labour will therefore have to expend effort for 6 hours a week (16 x 0.38) and this has a cost in terms of increased dietary intake. One approach to valuing this

> cost would be to take, for each of *dheki* and mill labour, the *incremental* energy expenditure associated with the respective tasks, in the *dheki* and in the mill; this requires knowledge of all the job changes, not just in rice processing but also changes elsewhere as a consequence of work opportunities/needs created by the movement of labour into processing work. Rather than trying in each case to estimate the marginal energy expenditure of rice processing a second, substantively identical, approach, which we adopt, is to value all the dietary energy costs above the base level in the rice mill case and in the *dheki* case on the basis that this gives a complete picture of the relative costs of two ways of doing a necessary task. The base level is defined as energy intake to support basic metabolic rate plus general domestic activity.

As it turns out, the adjustment for additional dietary energy use is substantial in the case of the *dheki* but is very small for the huller mill case; it consists only of the extra energy expended by two men each working 6 hours a week. The per hour calorie cost of *dheki* work is estimated (see pp.331-332) to be 100 and small-scale custom huller work is certainly no more than this. Rice has an average caloric value of 3465 per kilogram and, at an accounting price for rice of six takas per kilogram (see footnote 54), an extra 100 calories per hour for 16 hours a week gives a social cost per annum for each man of Tk 54 because of increased expenditure on dietary energy. Therefore we add Tk 54 to the efficiency wage of Tk 1095 derived earlier giving a complete efficiency wage for both the mill owner/operator and mill wage labour of Tk 1149. For future reference, we can note that the overall conversion factor from market to efficiency wages is $\frac{1149}{1513}$, equal to 0.76.

In addition to the labour used at the mill there is also one hour per maund of male family labour used in carrying the grain to and from the mill and half an hour per maund of female family labour used to clean the rice and separate the by-products. This male labour input is given an opportunity cost of zero at efficiency prices because farmers typically take their paddy for milling on local market days when they would anyway go to the local market for other purchases. In the Comilla project villages, market days in the 5-6 nearby markets were all on different days of the week and the farmer could choose to combine a trip to market and a trip to the mill on almost any day (if mills were very close to the house then it was normal to send children with small quantities of paddy every few days and if mills were a long way off farmers would take all their consumption paddy in one load; in either of these two cases the per maund transport costs are negligible.)

The female family labour input does have an opportunity cost; the derivation of this cost is dealt with in detail in the discussion below on dheki labour and the estimated social opportunity cost of this labour arrived at is 0.15 takas per hour. (For dheki labour there is an additional cost of 0.17 takas per hour due to the increased dietary energy utilisation associated with dheki work but for rice cleaning and by-product separation, largely a sedentary task, energy expenditure is not any greater than the base energy requirement for body maintenance and general domestic ability.) The opportunity cost of this female family labour, half an hour per maund for 3482.7 maunds (130 tons) at the rate of Tk 0.15 per hour, is therefore Tk 261.

FUEL: In 1980 the average price Bangladesh paid for imported diesel was Tk 16.80 per imperial gallon and the selling price at major centres in 1980 was Tk 22.50.45/ Allowing a notional ten per cent margin on the import price for local transport costs the conversion factor for diesel is 0.82. In fact only 77 per cent of the fuel costs were diesel (there were 2 electric mills) and some allowance should be made for any difference in the conversion factor for electricity. No estimate of the accounting price (the marginal social cost) for electricity is available and it is a complex thing to calculate. From the limited data available46/ it can be estimated that total revenue from electricity is only six per cent greater than the costs of fuel required to generate it. Taking into account all the costs of power station and distribution system construction and maintenance, wages and salaries etc it would appear that, at market prices, electricity is fairly heavily subsidised and that a conversion factor below unity is unlikely to be realistic. This view is confirmed by Bangladesh Power Development Board (1977, vol.I p.III F.82). Therefore, somewhat arbitrarily, a conversion factor of 1.5 has been used for the 23 per cent of power supplied by electricity which yields a composite fuel conversion factor of 0.98. At market prices fuel costs were Tk 3301 and at accounting prices they are Tk 3235.

SPARES AND REPAIRS: For this item covering a miscellany of industrial products we adopt Little and Mirrlees (1974, pp.217-219) approach of approximating the accounting price through use of a standard conversion factor (SCF). The SCF is a weighted average of the ratio of domestic prices to border prices (plus transport and trade margins). The standard conversion factor has been estimated in IBRD appraisal exercises to be 0.75 and this value has been used. Spares and Repairs at market prices are Tk 3197 and at accounting prices are therefore Tk 2398.

FIXED INVESTMENT COSTS: From the mill survey it is known that the fixed investment costs (Tk 11,095) are composed of construction (11.3%), installation (9.5%) machinery (74%) and other (5.2%) costs. Construction, installation and other costs include a high component of rural labour-directly and indirectly - and are converted at the shadow wage conversion factor (0.76);47/ the machinery costs are adjusted from market to economic prices at the SCF of 0.75. Together these yield an efficiency price of Tk 8350 for the investment of Tk 11,095 at market prices. This efficiency price has to be reduced by the discounted present value at accounting prices, of the scrap price for the mill after 15 years. The effect of including an allowance for scrap value reduces the fixed investment cost by about seven per cent48/ giving an efficiency price of Tk 7787.

Dheki Inputs

LABOUR: Seventeen per cent of the female labour used is hired and it has been inferred earlier that the social opportunity cost of this labour is zero. It will be argued later on that at full social pricing, when income distribution weights are considered, the 'cost' of hiring female labour is negative - the social worth of their consumption is greater than any output foregone by employing them. At this, the efficiency price, stage the argument is concerned only with the value of output foregone which is estimated to be zero, except for the opportunity cost of the extra food required to do the extra work.

Evidence from our survey on the structure of hired female earnings, given in more detail in Chapter Seven, shows that in Comilla women worked for wages on average for only 142 days a year and that 65 per cent of the days unemployed were because work was not available. Moreover, days worked actually means days with some work which in Comilla was, only exceptionally, more than about four hours work. It also includes time spent begging and gleaning responsible for nearly one-third of total income. Only a quarter of households with wage labour women owned any land and average land owned was 0.35 acres. Only a third of these households actually farmed and average land operated was 0.25 acres; it appears perfectly reasonable to assume therefore that no land-related productive employment was foregone by seeking dheki work.49/ Further, these wage labour women also came from smaller households than average and had much fewer assets than average - both reasons for discounting any possible cost in terms of foregoing productive non-wage work. The output foregone by employing these women can therefore be safely taken as zero but in the efficiency price some allowance must be made for the cost of the extra food they eat in order to perform the work. As described below this is estimated to be Tk 0.17 per hour. Out of 34,827 hours of dheki work, 5921 are undertaken by wage labour and the accounting price for this labour is therefore Tk 1007.

A non-zero marginal product of family female labour has been identified earlier (Section 6.7) as a prerequisite for any interest in the accounting prices of any other inputs; if family female labour has a zero social opportunity cost the dheki must be more socially cost-effective.50/ There is little direct evidence on this issue, the argument in favour of a small non-zero marginal product is largely inferential. Two pieces of direct evidence, apparently supportive, in fact tell us very little. First, the fact that the male marginal product is non-zero says very little about female marginal product since there is a very rigid gender division of rural labour. Secondly, the 17 per cent of dheki work given to wage labour suggests the family women employing this wage labour have, at that time at least, a marginal product more than equal to the wage. Only 14 per cent51/ of cultivating households employ wage labour; however, it is not because they are busy but because they are conscious of the social status associated with female withdrawal from physical work. For these women, the fact that they hire labour could quite plausibly be taken as evidence of their zero marginal product. This still leaves the other 86 per cent of cultivating households, who are neither the very poorest nor the richest rural households.

For these other households, it is almost certainly the case that their marginal product is below the wage rate, perhaps considerably below it, since they prefer not to hire labour which is readily available. However, even if their marginal product is only one-third of the wage, their social opportunity cost is greater than the efficiency cost of all the variable inputs to the huller mill.52/

There are two reasons for suggesting that a value of about one-third of the market wage may be plausible as an estimate of the shadow wage of family female wage labour. First, there are the consumption costs associated with the extra dietary energy required by the labour effort. Dheki work is long and tedious rather than arduous. It was shown in Section 6.5 that an average of about 45 minutes per day, or over 270 hours per year, is required per family - and there are over 13 million rural families in Bangladesh. Some crude estimate of the social cost of this work could be derived from a figure for the additional dietary energy requirements incurred. This is more complicated than it sounds and needs data which is not available. However, according to data compiled by Longhurst (1980, pp.158-160) on task-specific female energy requirements in Northern Nigeria, for rural social groups with similar body weights and almost identical activity patterns, the additional energy required on dheki work is probably between 58 and 150 calories per hour more than calories needed for the basic metabolic rate and general domestic activity.53/ This gives an average value of 104 which we round to 100 calories needed per hour of dheki work.

For each worker, this level of energy expenditure is equal to more than five per cent of daily dietary energy when dheki work, as commonly occurs, is shared by three women processing a maund of paddy in a day. (Also, it should be noted that in 1976 nearly 60 per cent of rural households were estimated to be deficient in calorie intake (Ahmad 1977, p.51); and per capita intakes are falling so the percentage of deficit households is rising (GOPRB 1983a, p.390).) With an average calorie intake per kilogram of rice of 3465, an accounting price per kilogram of six taka, the extra calorie need of 100 calories per hour costs just over 0.17 taka.54/ The dietary energy cost of the 28,906 family labour hours on dheki work is therefore Tk 5005.

Secondly, farm women are responsible, in addition to their domestic work, for a variety of non-field agricultural work. Chief amongst these are care and maintenance of all livestock and production in home gardens of fruits, spices and vegetables. There are few estimates available of productive hours per day by rural women for Bangladesh. A year-long survey (twenty-four hour recall once a month) for 57 married non-wage earning women from this study in Comilla and Tangail produced an average figure of 9.4 hours of productive work per day. 'Productive' work includes all domestic responsibilities as well as farm-yard and home garden work. Even if no allowance is made for meals, personal hygiene or leisure and assuming they sleep for eight hours these women are productively employed for nearly 60 per cent of their available time - under some very extreme assumptions.

If 60 per cent of the time spent in *dheki* work would otherwise be spent on productive work and if the marginal product of family female work time is only a quarter of the wage rate (at accounting prices equal to 0.9 x 1.08 or 0.97 takas per hour) the social opportunity cost per hour of *dheki* work (0.97 x 0.25 x 0.6) is 0.15 taka per hour. This would imply a social cost of Tk 4206 for the 28,906 family labour hours in *dheki* work. (No social value has been placed on the disutility of this labour effort and, at the very high levels of effort frequently expected of family farm women, incorporation of this factor could lead to a substantial increase in the social opportunity cost of family farm women's work; as with the huller mill estimates, this factor is not quantified but suggests the estimate of social opportunity costs is unlikely to be an overestimate.)

Adding the guestimates of the last two paragraphs gives a total social opportunity cost of Tk 9211 for family farm women in carrying out this *dheki* work. No special importance is attached here to these particular figures. The point being made is simply that there are a very large number of family female hours committed to *dheki* work and even very low estimates (totalling 0.32 Tk) of hourly social opportunity costs, based on plausible, probably conservative, assumptions, results in total costs which are large in relation to the other values involved in the *dheki*-huller mill comparison. These figures are used simply as a first approximation to the true social opportunity cost, at efficiency prices, for family female labour; the sensitivity of results to changes in these values are discussed later on.

A final point in support of some small positive opportunity cost for family female labour, though it is not quantified here, concerns the saving of retail margins through subsistence production. To the extent that family farm women are producing goods or providing services for self-consumption, i.e. subsistence goods, it can be argued that the additional direct consumption generated by their labour effort is greater in value than can be obtained by a hired labourer who has to pay for those same goods at local retail prices. This increases the opportunity cost of a unit of family labour, probably only by a rather small amount in Bangladesh, but is a further reason for believing that the social costs of family labour are not as low as sometimes presumed.

FIXED INVESTMENT: At market prices the cost of the 72 *dhekis* needed to provide a similar throughput capacity to one huller mill is Tk 9072. *Dhekis* are non-traded and the efficiency price is derived from the costs of wood and the costs of labour needed to make them. No information is available on these relative shares and the conversion factor used (0.78) is a simple average of the conversion factors for labour (0.76) derived earlier and an IBRD estimate for wood of 0.8.55/ This gives a dheki fixed investment cost at efficiency prices of Tk 7076.

Table 6.9 gives a summary of all these results at efficiency prices. It shows that if no account is taken of differences in the value of a unit of income earned by different income groups then the huller miller is more than three times as profitable in efficiency terms.

6.10 A COMPARISON AT SOCIAL PRICES

Two types of adjustment are usually made in converting efficiency prices to social prices. The first is to take account of the different social worth attached to an extra unit of current income going to the poor and going to the rich - the intra-generational adjustment. The second is to take account of the alleged sub-optimality of savings; i.e. the present value of the stream of consumption resulting over the years from a unit of invested social income is higher than the same unit being used for consumption today - the inter-generational adjustment.

A third consideration, the Accounting Price for Private Investment (APPI), is also relevant in this comparison. The reason is that the alternative use of the investment funds is not uncommitted social income but income of investors in dhekis and huller mills. These must be valued in terms of uncommitted social income.

The justification for the intra-generational (sometimes called interpersonal) adjustment lies in the assumption that aggregate consumption is an inadequate specification of the consumption objective unless explicit consideration is given to its distribution. This assumption is implicit in the income tax structures of most countries including Bangladesh. It is sometimes disputed whether SBCA applied to project selection is the proper vehicle for taking account of this distributional issue. Fiscal policy is an alternative. In countries like Bangladesh, however, the government is unambiguously incapable of operating such an alternative and if project selection criteria do not take account of intra-generational issues they are very likely to be neglected altogether - other than by those ministries whose business it is to operate public works programmes, rationing systems and other direct interventions all with known risks of corruption, waste and "trickle-up" and hence all implausible as redistributors.

The inter-generational adjustment values private consumption in terms of public savings. This is one of the most important contributions of SBCA methodology to the theory of public investment criteria. Sen (1972) has shown how main schools of thought on developing country public investment criteria differed most critically in their valuation of the costs associated with the generation of consumption through employment. Two common views were a) that it is equal to the wages bill, since resources so committed were almost all consumed and no longer available for investment; and b) that since labour frequently has a lower marginal product than the wage it receives it should be valued at its marginal product, (output foregone), which may be zero. SBCA essentially synthesises these two views, in a way Sen identifies, by separately assessing the value of output foregone and the costs of additional consumption incurred by creating a job and then adding them together to reach the shadow wage. The efficiency price of labour measures the output foregone. The social price takes account of the net costs to social income by also valuing the extra consumption generated relative to uncommitted social income. This approach has been operationalised in manuals of project appraisal to derive a value (s in Little and Mirrlees (1974), P_{inv} in UNIDO (1972), v in Squire and van der Tak (1975)) for uncommitted social income (essentially, public investment funds) in terms of consumption generated by employment. Using the terminology of Squire and van der Tak, $\frac{1}{v}$ is therefore the weight attached to consumption in terms of the numeraire, uncommitted social income.

Since, for the (non-OPEC) developing countries, savings ratios are low and capital is scarce, the present value of additional consumption induced by investing a unit of social income is generally estimated to be greater than the same unit consumed today. In other words, the social rate of return on investment is higher than the consumption rate of interest. In practice it has often proved quite difficult to estimate the correct weight for investment (Squire and van der Tak 1975, pp.104-109). (The theoretical basis for determining the optimum rate of savings is a subject of continuing controversy (see Sen 1984, pp. 11-18, 133-146 and 207-241) and there is not universal agreement on the view put here that market determined savings rates may be sub-optimal (Wellisz 1977). However, the view implicitly adopted here is not critical in the sense that alternative conclusions would emerge if a different view were to be taken on the use of a savings premium - this is not true though of the intra-generational adjustment, the use of consumption weights, which is the analytical focus of this section.) Appendix 6.1 discusses measurement of the savings weight for Bangladesh and demonstrates that the usual formula is inoperable since, with the negative growth rate of per capita consumption in Bangladesh, it yields a value of infinity for the weight. There is an alternative however; direct estimation of the weight is not required since a value is implicit in any estimate of _intra_-generational weights and this is the method used for our social prices.

Following Squire and van der Tak (1975) the _inter_-generational weight is labelled v and the _intra_-generational weight d; the net weight $\frac{d}{v}$ is labelled w. Use of these two weights provides a value for the benefits from the additional consumption out of wages; weights have to be estimated for female wage labourers, male wage labourers and for the consumption out of profits by mill owners. These benefits are deducted from the gross social costs of consumption in order to measure the net social cost of the additional consumption. (In theory, the benefits, measured by the value of milling services, should also be weighted according to the weight attached to the recipient of the benefits (the farm household); in practice this exercise is not needed because the benefits from each technique are identical.)

The gross costs of consumption are the values of social income foregone by committing resources to wages. In neither the _dheki_ nor the huller mill case are the gross social costs of labour equal to their accounting prices since they are not made from uncommitted social income but from the hirers of female wage labour and the employers of mill labour. This requires estimation of two further values; the weights, in terms of uncommitted social income, attached to the incomes of each of these employers. In the case of the _dheki_, wage payments are presumed to come out of consumption foregone by employing households. In the case of the huller mill, wage payments come out of profits and the weight for mill owners profits is therefore the same as that used for the mill owner as hirer of wage labour; this is a composite value based on the share of consumption and savings out of profits. The final two weights required are those used for converting the fixed investments from efficiency to social prices (the accounting prices for private investment for dhekis and huller mills).

To summarise, six weights are required, for hirers and providers of female wage labour, for hirers and providers of male wage labour and for investors in dhekis and huller mills. The following paragraphs describe the application of these weights in converting efficiency to social prices.

In Table 6.8, the wages for hired female labour are given as Tk 5329 and the consumption conversion factor is 1.08 giving gross consumption at accounting prices of Tk 5755; consumption foregone (in this case equal to the cost of extra dietary energy intake to perform the dheki work) is equal to the efficiency price of labour which is Tk 1007. Additional consumption is therefore Tk 4748. Therefore the total adjustment to the efficiency price of hired female labour, (Tk 1007) is equal to 4748 (W_{fh} - W_{fw}) takas where W_{fh} is the weight for employers and W_{fw} for providers of female wage labour. If W_{fw} is so much greater than W_{fh} that the value of dheki workers consumption is more than Tk 1007 the shadow wage will be negative - in other words there will be a net social benefit from committing resources to the consumption of female wage labour.

For mill workers, their additional consumption is equal to the wage less their marginal product; at accounting prices this is 1634 - 1149 = 485 takas. The net social cost of this additional consumption is equal to 485 (P_m - W_{mw}) where P_m is the value of a unit of social income committed to mill owners profits and W_{mw} is the value of the same unit committed to additional consumption by mill workers. P_m has been used on the basis that any change in mill workers wages will be reflected first of all in a similar absolute change in mill owners profits; P_m is defined more fully in the next three paragraphs on mill owners' profits. Application of this formula can also result in negative social costs of labour but with an efficiency price of labour of Tk 1149 this will only occur if W_{mw} is greater than P_m by more than 2.4 since the full shadow wage is equal to takas (1149 + 485 (P_m - W_{mw}).

The mill owners rate of return, computed from Table 6.8, is nearly twenty per cent, i.e. an income of Tk 2171, (value added plus the notional wage). The estimate of his additional consumption as a result of his investment requires a knowledge of what his rate of return would have been if he had invested elsewhere (Little and Mirrlees 1974, p.200). This is not available but the results are not very sensitive to the precise value used so we have assumed a value of 15 per cent which is slightly higher than can be obtained from a term deposit with the nationalised banks. (Strictly speaking allowance should be made at this stage for depreciation, tax payments and loan repayment but the mill owner will typically not make an allowance for depreciation, will evade any taxes due and generally (see Appendix 6.1) will not have borrowed to raise the investment funds). The additional profits from milling therefore come to Tk 507. The social value of this income will differ according to whether it is used for consumption or for savings. Again, no direct evidence is available on this point; total monetised savings were only about 2 per cent of GDP in 1978 (IBRD 1978, p.43) but, clearly, entrepreneurs who invest in these mills will have high savings rates compared to average; the conservative guesstimate that we use is 20 per cent which yields Tk 101 in savings and Tk 406 in consumption out of mill owners profits. The consumption is converted into accounting prices by multiplying by the consumption conversion factor (1.08) to give a value of Tk 438.

The social benefits from mill owners own consumption (Tk 438) - is 438 W_{mh} where W_{mh} is the weight attached to their additional consumption. Mill owners' savings, 507 - 406 = 101 takas, are multiplied by the owners APPI to estimate their social worth. Their APPI depends upon the value, in terms of the numeraire, of the alternative use of their funds. Evidence on the sources of funds which include sale of land and gold, savings, loans and other enterprises, is used to derive this value labelled $APPI_o$. This value is also used to adjust the efficiency price of their fixed investment to social prices.

The net social cost of mill owners additional profits is therefore 438 $(1 - W_{mh})$ - 101 $(APPI_o)$. This has to be added to mill owner labour costs at efficiency prices to derive the net social cost of the mill owners' labour, management and capital. To estimate Pm, the gross social value of a unit of mill owners' profits is required which is given by the weighted average of W_{mh} and $APPI_o$. These weights correspond respectively to their marginal propensity at efficiency prices to consume and to save and are 0.81 for W_{mh} and 0.19 for $APPI_o$.

Finally, both the dheki and huller mill investments (Table 6.9) have to be weighted by the relevant APPIs to arrive at the social opportunity cost of the investments. In the case of the huller mill this is Tk 7787 $APPI_o$. In the case of the dheki however W_{fh}, which measures the social value of consumption by hirers of female wage labour, will not be the correct income weight for the investors in dhekis; W_{fh} depends upon the distribution of the employers of female labour and these households are likely to be richer on average than the households that buy dhekis who are much larger in number. Therefore, though the same assumption is made as for W_{fh}, that the alternative use of funds is for consumption, the weight for dheki investors W_{di} is separately calculated. At efficiency prices (Table 6.9) dhekis cost Tk 7076 and the net social cost of the investment is therefore Tk 7076 W_{di}.

The derivation of W_{di}, W_{fh}, W_{fw}, W_{mw}, W_{mh}, and $APPI_o$ is given in detail in Appendix 6.1. They are estimated to be:

$$W_{di} = 0.75$$

$$W_{fh} = 0.17$$

$$W_{fw} = 2.50$$

$$W_{mw} = 0.75$$

$$W_{mh} = 0.11$$

$$APPI_o = 0.41$$

and

$$P_m = 0.17\ (= 0.81\ W_{mh} + 0.19\ APPI_o)$$

These values have been applied in the ways described above to the results at efficiency prices to arrive at the social price comparison summarised in Table 6.10. It is apparent that a radical change in relative profitability has occurred. At efficiency prices the huller mill has a net present value five times greater than the same throughput capacity supplied via the dheki. At social prices the dheki has a net present value nearly five times that of the huller mill. Both techniques have enhanced net present values: the huller mill by Tk 4138 and the dheki by Tk 77,119.56/ Some comment is required.

It could be suggested that the results at social prices are themselves sufficient evidence of the inappropriateness of introducing distributional considerations into the analysis. More specifically, the results seem to make nonsense of basic economic concepts; for example, at social prices the dheki is more capital intensive (higher capital-output ratio) than the huller mill and even though the dheki uses 60 times more labour its wages bill is less than for the huller mill - indeed, it is negative.

There are two types of answer to this charge. First, that all the value judgements have been explicitly stated and that the two to which the results are most sensitive b, the critical consumption level and e, the elasticity of the marginal utility of consumption with respect to consumption (See Appendix 6.1), have been derived reasonably systematically and are based on empirical findings. Moreover, stopping the analysis at efficiency prices also implies either (a) the false judgement that GOPRB can implement redistributive measures via taxes and transfer, or (b) value judgements about distribution just as strong as we make above, viz. that all forms of current income are equally valuable and that future consumption is worth less than current consumption but differences in the levels of current consumption amongst consumers now are irrelevant. Such values are not easy to support since they are inconsistent with government preferences, demonstrated for example through income tax structure and other aspects of fiscal policy. Nevertheless, the effects of alternative value judgements are described below.

Secondly, it has to be remembered that the application of SBCA here is not to an ex-ante evaluation of alternative public sector investment proposals, but to an evaluation of private sector behaviour in a rural economy where massive differences in access to economic resources is known to be resulting in deepening poverty for many households already very poor. In these circumstances the question that the analysis is addressed to is not one of 'selection or rejection' but of intervention to correct the undesirable consequences of market behaviour. The results do not automatically lead to a conclusion that the huller mills should be somehow stopped (cf. Sen 1984, pp.224-241). They suggest, rather, that the social costs of effectively implementing such a strategy should be compared to the costs of alternative interventions. If the government has other programmes which are designed to improve the consumption level of those households dependent on female wage labour income then it would be consistent to use a lower value of d, the intra-generational weight, for these women which, if low enough, would result in the huller mill being socially preferred. The advantage conferred by use of social prices lies in the identification of what the objectives of such interventions should be.

6.11 SENSITIVITY ANALYSIS

Inspection of Tables 6.9 and 6.10 clearly suggests that the single most important value affecting the relative social profitability of the two techniques is the size of the benefits (negative social costs) associated with employment of female wage labour. The implied value for d_{fw}, the weight on their consumption relative to average consumption, is 2.85.57/ This value would be reduced if either b or e was lower. It is improbable that any lower value for b than that used, 80% of a tightly-defined dietary norm, could be substituted and still retain a satisfactory interpretation as a 'critical' consumption level; given that the consumption level of female wage labourers is only 63% of b the change required to reverse the conclusion would have to be a much lower value.

The value of e is much more open to question. However, it can be demonstrated58/ that the switching value for e is between 0.25 and 0.50. In other words, there would have to be a substantial change in assumptions concerning the relative weight attached to different current consumption levels in order to reinstate the huller mill as the more profitable technique.

The value for the opportunity cost of female family labour was earlier mentioned as subject to some uncertainty. Any revision is likely to be downwards and the effect of this would be to increase the NPV of the dheki so the conclusions are insensitive to the imputed values for family female wages that have been used. This is also true for the value of output which has been considered equal in the two techniques but where there was some evidence (Section 6.4) that output from the huller mill could be marginally less valuable because of broken grains; adjustment for this evidence would result in an increase in the relative profitability of the dheki.

Two other values concerning which there was some uncertainty were the unit cost of the dheki and its capacity utilisation. However, it is evident from the benefit-cost ratio of 15:1 that even quite large increases in these dheki investment costs would not change the social profitability rankings: in fact, any correction for underestimation of unit costs and overestimation of unit capacity will only make the dheki less socially profitable than the huller mill if the correction requires an increase by a factor of 2.6 in the dheki investment costs.

Inspection of Table 6.10 indicates that any increase in the proportion of hired female labour hours, above the 17 per cent of total hours assumed, must increase the social profitability of the dheki. A downward revision in the share of hired labour in total labour would have to be very substantial to reverse the ranking of the two techniques however. If the share of hired labour in total dheki labour was only ten per cent, the dheki would still have a benefit-cost ratio fifty per cent greater than that of the huller. As the discussion in section 6.2 suggests, error in this value may be up or down, but either way it is not likely to be more than a few per cent.

The results are particularly sensitive to only one other value: this is W_{fh}, the weight attached to consumption by households that employ dheki labour. A value of 0.17 was derived on the basis that these were the fourteen per cent richest rural households. If these households were more evenly distributed this weight would increase and, therefore, the net social benefits from employing female wage labour would be reduced. But the switching value for W_{fh} is approximately 2.09 and would imply that the households that employed wage labour for dheki work had incomes only 65 per cent of average rural incomes. This is clearly unlikely. Obviously, combinations of reductions in the value of e and increases in the value of W_{fh} would reduce the magnitude of the changes required in either one to make the huller mill more profitable. But these changes would still be quite large. If e is reduced by a factor of four to 0.5 the switching value of W_{fh} is about 0.8; i.e. even with e = 0.5, hirers of dheki labour would have to have incomes only twelve per cent above the critical consumption level (see Appendix 6.1) in order to induce the switch in results. This income level would be above the average by less than five per cent which is unlikely. It is apparent from these switching values that the greater profitability of the dheki is a robust result - given the use of income distribution weights.

NOTES

1. One difference lies in the lower yields of head rice (unbroken rice) associated with dehusking and polishing raw rather than parboiled paddy. This difference applies to both the traditional methods and to the mechanical rice mill but we have no evidence to establish whether comparative performance of these two traditional methods and the mill in processing raw paddy is different to the comparative performance of dheki and the mill for parboiled paddy.

2. Households are grouped physically into baris which usually have three or more households of families all related by blood or marriage.

3. Note that the decile groups are on a household basis, including non-rice farmers, whereas these percentages are based on rice farmers; this is why dividing the number of cases by proportions hiring in will yield different totals in each decile group.

4. In practice, as with female wage labour use, some farms may only use the mill at harvest time.

5. The mills surveyed varied from 14 years to just a couple of months old. In several cases the mill was purchased second-hand and the age of the mill was not precisely establishable. On variation in engine/motor size, torque readings have shown (M. Lockwood, personal communication, November 1983) that 5 horse power is sufficient for the most commonly found huller size (the number 8). The universal practice of using excessively large power sources probably reflects experience of maintenance problems - larger engines working inefficiently will still do the job.

6. In a very few poor households wheat is prepared for consumption without mill grinding, but it is poverty not preference that dictates this consumption habit.

7. In Tangail in the wet season roads were impassable for jeeps and rickshaws and villagers only five miles off the main road were effectively cut off from rice mills. In Comilla in the dry season (when country boats could not be used) villages were similarly affected.

8. Assuming that 60% of mills are unlicensed (the approximate sample survey result); that the average of 130 tons per year milled in the six Comilla mills monitored weekly for over a year (see section 6.6) under-represents, because it is an important wheat area, by twenty per cent the quantity of paddy processed annually in the 'average' small mill; that 15 per cent of production is marketed surplus but that only twenty per cent of this is processed in the small custom mills (as opposed to the major mills used by the Food Department and many traders); then, given the official Food Department estimate of 10,193 small mills in 1980 and production net of seed and losses etc (10%) and sale (15%) of 15.5 million tons, about 30% of farm-level consumption paddy was mechanically milled in 1980. The range given is this figure plus or minus five per cent.

9. There are very rare cases where women will work at mill premises to separate the by-products (husk, bran and broken grains mixed in with them) but these women work for

the customer not the miller. It is much more common for the by-products to be taken back to the farm for separation.

10. Mr Rizvi, loan manager Shilpa Bank, Dhaka, in an interview with the author in 1979. An instructive note on investment appraisal methods for rice mill proposals for Bangladesh (Harriss, 1978 Appendix A) deals with other aspects of appraisal methods.

11. One problem on the experimental design is deciding a) what level of polish to set for the laboratory mill, and b) having decided, being able to maintain that consistently across samples of varying quality. In the results the average level of polish measured by the percentage difference between brown rice and polished rice yield was exactly 6% with a range of 3.1% to 7.8% and a standard deviation of 1.08. No precise figure for per cent of polish is prescribed for rice millers in Bangladesh but 6% is the level specified for mills working on government contract in India.

12. Unfortunately, results on loss of solids during cooking were not obtained for the laboratory-model modern milling samples.

13. This is a very significant practice: Table 6.2 shows that total loss of solids, without any broken rice, is greater than 7 per cent which is more than total food losses in the whole postharvest system from cutting up to milling. Clearly, recovery of this loss of solids is much more valuable than any other loss-avoidance practice. An alternative to using the water afterwards would be to use less water in the first place and cook the rice until it is dry. However, use of excess water is universally practised in Bangladesh villages because otherwise constant supervision is required and because with the types of fuels available temperature control is difficult.

14. Personal communication from Ruth Dixon (June 1980).

15. In some cases the dheki is used to put a final polish on the rice after it has come back from the mill. In many cases, farm women continue to dehusk and polish some paddy, especially in slack seasons and almost all households (excepting perhaps those with more than one mill on their doorstep) would maintain a dheki, or access to one, in case of emergency.

16. There is an enormous variation in daily wage rates and they are almost universally paid in kind - meals and paddy or rice. Using an approximate average value of 2 meals (Tk 2.5 each) and 1 seer (0.93 kilograms) of rice (Tk 150 per maund) the wage rate per maund (37.32 kilograms) of paddy husked comes to a figure of Tk8.75; this is a slight

underestimate because bran and husk kind payments, which were common, are excluded and to allow for these payments the wage rate per maund has been increased by Tk 0.25 to Tk 9. However, since the quantities processed each day were normally less than one maund, daily earnings were generally lower than this; in Comilla they averaged Tk3 plus meals.

17. The initial survey included investment costs, hourly and daily throughput, labour use, maintenance and repair costs; however, data obtained from the subsequent weekly survey of eleven mills has generally been used in this analysis since it could reasonably be assumed to be much more accurate.

18. A piston failure led to the flywheel breaking, spectacularly, into 3 parts (one of which went through the roof and several yards across the compound into a pond and another of which injured the operator). The owner-operator managed to obtain two-thirds of the spares costs by selling the broken parts for scrap and made the repairs himself.

19. This mill had the smallest diesel engine (8.5 H.P.) of any in the survey and the most remote location; relatively high charges (Tk6 paddy, Tk12 wheat) relatively high capacity utilisation (38 per cent), and very few days of non-operation (one period of 28 days and one of 3 days in 391 days) together contributed to this high level of profit.

20. Casual inspection of Table 6.5 does not confirm this since four of the five electrically powered mills (Mill nos. 5,6,7 and 11) were in the top four positions with respect to fixed investment costs. However, these mills, unlike the diesel mills, were mainly recent investments (three of them during the survey period in 1979); also the electrical motors purchased most recently (30 HP at an average price of Tk19,000) were in excess of current power needs and had been purchased in anticipation of subsequent expansion to a line-shaft operation. Harriss (1979) and the Bangladesh Power Development Board (1977) also gives figures which suggest that investment costs are greater for diesel than electrical power.

21. Note that: a) Electrified areas can (and inevitably will) attract investment funds from non-electrified areas b) investment planning horizons are short with high but variable rates of return c) transformation from diesel to electric mill investments are common and aided by the effective second (n)-hand market, human capital assets in operating milling equipment and goodwill inherited, even when locations change.

22. There is a strong case for carefully controlling the success of rice mill electrification applications. As

described above, rice milling is easily the most important load demand in the rural electrification project in Bangladesh. A combination of many relatively high load factors but low and irregular power demand can substantially increase the social costs of electricity (greater investment, more complex management and low reliability). There is some evidence from Madhupur that this is an emerging problem for the REB; the number of rice mills in the thana town increased from five to twenty immediately upon electrification; 63 per cent of non-operational days were lost due to power supply failure. (See the section on fuel costs below.)

23. We have no direct evidence on this issue and hesitate to state a definite position; continuing dependence, at the margin, in Bangladesh on thermal generation of electricity - and the special low-charge rates used for rural small industries, together with the continuing high transmission losses in the distribution network suggest that market prices may be biased against diesel fuel.

24. It has been observed that it takes 10 woman hours to dehusk and polish rice from one maund of paddy using a _dheki_. The average throughput in a mill run by two men is 11.78 maunds (say 12) per hour (see below). If no account is taken of relative labour power invested in the different genders (an unimportant omission) or of the differences between paid labour time and actual mill operating time (an important but not crucial omission) it can be easily calculated that the productivity of mill labour is sixty times greater than _dheki_ labour.

25. The benefits, in principle, include both consumption and savings accruing to different income groups including the government as an income group; government income in fact is barely relevant here since the only direct effect on its income is for mill licenses which are small and often unpaid.

26. These were available as a collection of eight pages summarising accounting prices from 13 appraisal reports prepared from 1975 to 1981 by the World Bank. No fuller reference (other than project name) is available. All future references to World Bank estimates are to these figures.

27. It is probable, in comparing two techniques, that the technique with lower investment costs also has the higher operating costs. The exact choice of discount rate will only matter if, at a rough value of the ARI, the difference between the life-time operating costs of the two

techniques is fairly close to the difference in investment costs. As the analysis shows, at none of market, efficiency or social prices is this the case.

28. Switching values indicate the change in value of a key parameter that it is necessary to make in order to reverse the choice of technique derived using the original value.

29. In addition to the two best-known texts (UNIDO (1972) and Little and Mirrlees (1974)), Lal (1972), Squire and van de Tak (1975), Scott, MacArthur and Newberry (1976), UNIDO (1978) and Irvin (1978) are important; the last two are respectively the best practical guides to and the best review and synthesis of SBCA methods.

30. Social analysis should be based upon the probable future mix of power sources. The mix (1/3:2/3 adopted here) is certainly an over-estimate of the current importance of electric mills but they can be expected to grow in relative importance and an estimate of one-third electric is probably a reasonable approximation of the medium term future distribution. The limited differences, according to survey results discussed above and summarised in table 6.6, between diesel and electric mill performance suggest that more refined analysis of this issue would not affect the thrust of the results.

31. The one important argument against this would be the charge that in other districts the wheat share in total output is lower and in some, e.g. Patuakhali, insignificant for agro-ecological reasons. This is valid in that Comilla has been consistently one of the leading wheat producing districts; it would suggest that, if wheat grinding at huller mill sites is more socially valuable per unit operating time than paddy milling, the case for the huller has been underestimated since the two are complementary investments.

32. The average figures for throughput per hour by huller size were: large (no 2) 13.69 maunds; medium (no.8) 12.65 maunds; and small (no.4) 7.84 maunds. Whilst these results are based on precise measurements using watches and scales, it has to be emphasised that efficiency, relating mainly to operator capability, equipment maintenance and, to some extent, demand management, is subject to wide variability.

33. This estimate is very similar to that which can be derived from the Rural Electrification Feasibility Study (Bangladesh Power Development Board 1977). A 15 horse power electric motor is priced at Tk15,600 and a 20 horse power diesel engine is priced at Tk20,000. The wholesale price

index for industrial goods has increased from 431.59 in 1977/78 to 604.73 in 1980/81, (GOPRB, 1983 P335) an increase of 40 per cent. The electric-diesel share is 1/3:2/3 and the share of paddy in investment costs is 0.38. The cost of a medium-sized huller No (8) at 1980 prices is Tk1541 giving a total investment cost of Tk11,400. Their estimate is based on higher real prices for electric motors than our sample farmers paid but makes no allowance for installation costs.

34. There is an implicit presumption that the rates of profit in the two activities are equal; i.e. that the unit throughput charges are in proportion to processing times required. If this were true then the ratio of wheat to paddy charges divided by the ratio of paddy to wheat maunds per hour would equal one. In fact, using the average throughput figures quoted in the text this value is equal to 0.9. (2.65 ÷ 3.04) Paddy, on average, is slightly more profitable but the result is regarded as sufficiently close to ignore differences in profitability of paddy and wheat operations.

35. Nearly 2/5 of households don't own a dheki. These are mainly landless households. To the extent that the rice they consume is dheki processed the average utilisation of dhekis will be greater than 1.44 tons; this may be true because of kind payments in the form of meals or simply through borrowing the dheki to process paddy. There appeared to be no satisfactory way of adequately allowing for this but it is probably too important to be simply ignored. Arbitrarily, it was assumed that these households were enjoying 80 per cent of average consumption levels and that half of their consumption rice was dheki processed. Effectively, this means that only 3/5 of households process 1.44 tons on their own dhekis; the other 2/5 have processed for them an average of 0.56 tons on others' dhekis. The average throughput per dheki then becomes 1.81 tons [1.44 + (0.56 x $\frac{2}{3}$)].

36. By completely ignoring whatever little marketed surplus is dheki-processed there is some tendency to over-estimate the number of dhekis displaced; on the other hand, average intakes actually consist of a few households that consume well above average and many that consume below average thus under-estimating the number of dhekis. The sensitivity analysis considers the implications of changes in the numbers of dhekis but fine-tuning of results to take account of these sorts of considerations does not affect the thrust of the evidence.

37. As with the cost of a shelter for the dheki the costs of maintenance are ignored. These costs are negligible and given the uncertainty regarding the total investment costs, adjustments to allow for them would be meaningless.

38. A major cost item of an engine is the fly-wheel and one of the Comilla mills recovered two-thirds of the cost of a new fly-wheel by selling the three broken pieces as scrap-metal in Dhaka.

39. To quote Little and Mirrlees (1974 p.270) precisely it is "the value of uncommitted government income measured in terms of consumption committed through employment".'

40. Lipton (1983b) develops the concept of 'transformation capacity' to analyse the relationships between household labour power and labour income.

41. Of the various figures available on labour use in agriculture the most studied and the most reliable were for the cereals. Since cereals (rice mainly, wheat and several minor cereals) account for well over 70 per cent of value added in crop production it was decided to use data on cereals to estimate labour productivity. The estimate of value added (value of output less purchased inputs except labour) is from GOPRB (1980b, pp.618-621 and 1983a, p.176).

42. Actually, the marginal social value product should have been calculated at accounting prices which is an apparently serious omission. However, on reflection, it does not matter much, at this stage. The accounting value of crop output is dominated by the accounting price given to paddy. The value of consumption which is dominated by rice, depends most critically upon the accounting price for rice. And, since the accounting prices for paddy and rice bear more or less identical ratios to their respective market prices no great harm is done by not making the adjustment to accounting prices before estimating marginal productivity and instead using the consumption conversion factor to derive the accounting price.

43. In the manually-operated tubewell project from the source cited in footnote 26. The high weight is because village prices of paddy considerably underestimate (in this calculation by 41 per cent) the costs of paddy procured by the government. This is consistent with other studies (e.g. Scott, MacArthur and Newberry 1976) where agricultural products have an accounting ratio greater than one. In general, foreign exchange saved through domestic production of a good that is also imported and subsidised will have an accounting ratio greater than one.

44. The mill workers annual wages of Tk 3981, including kind payments (meals especially but also clothes) which are quite considerable are similar to the earnings of a permanent farm worker and provide no evidence of a skill premium. The social value of the disutility of labour effort, given the low marginal product of labour, is not something which could be of any great significance in the Bangladesh rural economy.

45. Prices from GOPRB (1983a, pp.458 and 475) using a 1980 taka-dollar exchange rate of 16. This is the official rate which, with the use of conversion factors to 'border' rupees in Little and Mirrlees terminology, is appropriate in this context.

46. GOPRB (1983a, p.424) gives fuel costs (1980-81) of Tk 1361 million and gross collection from billing of Tk 1456 million.

47. This conversion factor was calculated for mill wage labour on the assumption that output foregone was in agriculture and that mill work involved about 100 calories per hour extra energy. By using the same conversion factor here the same assumptions are made; there is no direct information available on the labour used in mill construction but we are reasonably confident that no great error is introduced by using this conversion factor.

48. Scrap value has been calculated on machinery only for which the accounting price is Tk 6158. It has been noted (see footnote 38) that scrap values may be very high relative to investment costs; we assume that fifty per cent of the machinery investment cost will be recovered. Recovery is effected 15 years after investment and is therefore discounted by a factor of 5.47 giving a scrap value of Tk 563 equal to 6.7 per cent of total investment costs.

49. The fact that households operating were larger in number than households owning land is surprising. However, the sample size here is only 16.

50. There is one (plausible?) exception to this. If at social prices the accounting price for private investment of mill owners is zero, if they earn only an average rate of return, if mill workers are very poor and the costs of fuel and spares are low it is possible that the social costs of mills will be less than the social value of consumption foregone by farm households investing in dhekis. This investment is undiscounted (it occurs in year 0) and if the income level of the households investing in dhekis is at or below the base consumption level a relatively small elasticity with respect to income of the marginal utility of

income would put a high social premium on their consumption foregone.

51. According to the village census in Comilla approximately 10 per cent of all households hire wage labour women and approximately 76 per cent of all households cultivate rice. With Comilla's better developed infrastructure (more mills) and lower inequality in land distribution (fewer big farmers) these results probably understate the average. In Tangail, 9 per cent of households hired women and only 62 per cent of households were cultivators. The average percentage of cultivators who hired women in the two districts is just under 14 per cent.

52. 28,906 family female hours at one third the hourly rate of 0.9 takas yields, with a consumption conversion factor of 1.08, a social opportunity cost of Tk 9365 compared to Tk 8192 for labour, fuel and spares and repairs with the huller mill.

53. According to Longhurst (1980, pp.158-160) body maintenance needs are estimated to be 1.5 times BMR (basic metabolic rate) or 1890 calories. This is about 78 calories per hour. General domestic work is estimated to add between 30 and 150 calories per hour to this need; on average, 168 calories per hour are needed. Husking rice requires between 148 and 240 in excess of BMR x 1.5 and therefore requires between 226 and 318 calories per hour in total. Therefore the additional energy required because of _dheki_ work is between 58 and 150 calories per hour in total and averages 104.

54. The calorie content of rice is from GOPRB (1983a, p.580) and the accounting price for rice is that given by the World Bank for 1980 of Tk 5968 per ton.

55. This was the figure used in the mangrove afforestation project appraisal in 1980. An alternative mentioned is a much higher value based on the extra paddy output if wood replaces manure as a fuel thereby releasing the manure for use as fertiliser. This estimate of the conversion factor is 2 and on the assumption that wood accounts for 50 per cent of _dheki_ costs the conversion factor for the _dheki_ from market to efficiency prices would be 1.4. This would yield a fixed investment cost of Tk 12,700 for the _dheki_ compared to Tk 7,787 for the mill at efficiency prices.

56. The fact that both techniques yield a positive NPV could be literally interpreted as sanction for both to be undertaken which is clearly wrong. It implies that the accounting price for output is too high or that the ARI is

too low; either of these can be adjusted so that only the *dheki* technique has a positive NPV.

57. The average consumption level is Tk 1791 (see Appendix 6.2) which gives a value for d of $(1791 \div 1061)^2 =$ 2.85. This is consistent with v at the average consumption level which is 1.14 since 2.85 - 1.14 = 2.50 = W_{fw}.

58. From the earlier analysis the income levels per consumption unit of the groups affected by the choice of techniques are given or can be inferred. They are female wage labourers Tk1061, female family investors and mill wage workers Tk 1939, families hiring female labourers Tk 4072, and mill owners Tk 5062. Using these income levels and assuming all other values, including $APPI_O$, are constant an e of 0.50 makes the NPV of the huller mill about Tk 5000 less than the *dheki* but at e = 0.25 the huller has a higher NPV of about Tk 4000.

Appendix 6.1

Derivation of Distributional Weights

Introduction

This appendix provides estimates of the distributional weights(w) on consumption for four consumer groups; female wage labour (W_{fw}); male wage labour (W_{mw}); hirers of female labour (W_{fh}); and, hirers of male labour (W_{mh}). The Ws are composite weights consisting of d, the intragenerational weight which measures the value of a consumption group's incremental consumption relative to incremental consumption at the average consumption level, and v, the value at the average consumption level of uncommitted social income (the numeraire) in terms of consumption. Any given consumption level at accounting prices is multiplied by d and by $\frac{1}{v}$ to estimate its social worth ($W=\frac{d}{v}$). This is then subtracted from its social cost to obtain the net social opportunity cost of additional consumption. Adding this result to the efficiency price gives the total social cost of employment.

The appendix also estimates the Accounting Price for Private Investment (APPI) by mill owners, where it is labelled $APPI_o$, and dheki owners where the APPI is estimated as W_{di}. These values are used to estimate the gross social costs, in terms of uncommitted social income, of the fixed investments.

The values for W are derived from assumptions concerning the elasticity of the marginal utility of consumption with respect to increases in consumption, (e), and estimates of the critical consumption level (equal to b, the base consumption level in Little and Mirrlees's terminology). b is defined as that consumption level at which an extra unit of consumption is equally as valuable as a unit of uncommitted social income. At the average consumption level d=1 but at the critical consumption level (b) $\frac{d}{v} = 1$

The value of d varies with consumption levels but v is constant and estimates of W therefore contain an inferred value for v. v can also be independently estimated and depends upon the marginal social productivity of capital, the marginal propensities to consume and save and the consumption rate of interest (CRI). Both the marginal social productivity of capital and the CRI are in part dependent upon the value of e and in a fully articulated economic model the independent and inferred values of v should be consistent. Little and Mirrlees (1974, chapter

XIII) in fact give two ways to estimate v (s in their terminology) directly. However, for Bangladesh, neither of the direct estimates based on their methods yield values of v which are consistent with that inferred from e and b. The inferred, consistent (and lower) value of v is used in estimating the social prices in the text. However, the interdependence of the variables on which d and v are based is, even so, not fully respected and some inconsistencies remain (see below). In practice, it is difficult to achieve complete consistency as even the most careful studies have recognized (Scott, MacArthur and Newbery 1976, pp.45-48); the synthesis of SBCA with regional or national planning models has not yet achieved the combination of analytical rigour and empirical soundness (there are data availability and accuracy problems) to ensure consistency. But, the best is often the enemy of the good and provided the weights used are an improvement on no weights then they are worth using.

Two further introductory points need making. First, in an ideal application of SBCA these weights, or values such as b and e to calculate them, would be available, specified by the Planning Ministry of Bangladesh; these are what UNIDO (1972) call National Parameters. Since they are not available, and since the results are very sensitive to the assumptions made, the derivation and application of the weights given here must be treated with caution. The conclusions to be drawn are only relevant if the assumptions made correspond to the value judgements that a (benevolent) government would in fact endorse. Whilst sensitivity analysis can show which assumptions are critical to the conclusions there is no substitute for consistency, between appraisals of different projects, on values of key variables. In this study, the only such variable on which there seemed to be evidence and agreement was the ARI.

Secondly, the notation used is that of Squire and van der Tak (1975) and Hansen (1978) and the methods used, like theirs, are Little and Mirrlees methods. A justification for methods used here is not provided since these authors' works and Irvin (1978) provide excellent and well-known explanations of the operating principles. (With a few exceptions, largely in detail, UNIDO (1972) provide a similar explanation).

The only main departure here from Little and Mirrlees methods is in the estimation of the critical consumption level(b). According to them (Little and Mirrlees 1974, pp.244-5) this can be taken as the mid-point between the lowest income at which people succeed in living and the minimum tax level in Bangladesh; such an estimate is dependent upon derivation from a fiscal policy which is rather crude and operationally weak. Other methods they suggest (notably through examination of consumption subsides) also involve difficult inferences from government policy which it is felt best to try and avoid. Such methods are likely to give extremely high estimates of b. The alternative used here is to employ a Food Adequacy Standard (FAS). The essence of the approach is based on ideas developed by Lipton (1983a); he has demonstrated that, until recently at least, estimated poverty lines considerably overstate the numbers of people at serious risk of suffering malnutrition. World Bank identified poverty lines are re-estimated here to distinguish a critical income level that separates the ultra-poor from the poor and from the rest. This is calculated as the consumption level at which eighty per cent of the dietary norm is just met (see section on the critical income level below). A nutritionally determined critical consumption level allows an easier intercourse with policies expressed in terms of basic needs (and/or 'well defined' poverty line estimates) which is a very significant advantage in terms of policy interpretation for planners.

The rest of this appendix is in four sections devoted respectively to the elasticity of the marginal utility of consumption with respect to increases in consumption (e); the critical consumption level (b); the premium on savings (v); and, the accounting price for private investment. W_{fw}, W_{mw} and W_{mh} are given values in the second of these parts but their justification depends upon arguments in the next sections also. $APPI_o$ W_{di} and W_{fh} are estimated in the final section.

A. The elasticity of the marginal utility of consumption (e): This measures the percentage change in the marginal utility of consumption from a one per cent change in consumption. (Since utility falls as consumption rises the negative value is used; note also that it is assumed all income is consumed so the terms income and consumption can be used interchangeably). It is not proposed to dwell here upon the validity and interpretation of this concept which has been dealt with at length in the literature; (see eg. Irvin 1978, pp.139-145 and references therein and, more recently, the review by Ray (1984) which provides a detailed discussion of alternative weighting systems in the framework of benefit-cost analysis and essentially supports a system of variable weights derived from a constant elasticity for the marginal utility of consumption as used here - non-marginal income changes and changes in weights over time are two further issues, ignored here, which Ray explores in some detail.) The value of e is the sole determinant of the relative weight attached to income increments at different current income levels which are equal to $(\frac{b}{c})^e$ where c is the income level in question. It is also a determinant of the savings weight (v) because v depends upon the Consumption Rate of Interest (CRI) which is defined as e times the rate of growth of per capita consumption. (Various alternative formulae exist to allow for impatience and for the fact that in future, with population growth, a given level of per capita consumption actually implies increased aggregate welfare - neither are incorporated here). At efficiency prices the implicit value of e is zero.

There is considerable doubt about the feasibility of accurately determining a value for e. Little and Mirrlees (1974, p.240) suggest 'that most people would put 'e' in the range 1-3'. Noting the radical effects on project choice using high values of e, Hansen (1978, p.73) suggests a value of 0.5. Squire and van der Tak (1975, p.103) recommend a value of 1 with a sensitivity analysis for values of 0.5 to 1.5 or possibly 2. Ray (1984) p93 states that 'those using the constant elasticity form tend to set e in the range of 1-3, with most preferring values on the low side'. Methods for deriving e based upon an economy's tax structure assume that the implied governmental estimate correctly reflects the societal value. A second group of methods based on consumer behaviour (eg. Lal 1972, pp.150-152) face formidable data problems.

Chowdhury (1981) has applied both methods to Bangladesh to obtain estimates, respectively, of 1.79 and 2.54. Lal (1972) using the second method for India obtained a value of 2.3. Following Little and Mirrlees (1974, p.241) the approach adopted here is not to attempt direct estimation of e but instead to assign income weights - with implicit values of e - to broad groups of consumers. The weights adopted are consistent with a value of 2 for e. The essential defence of the implicit value for e of 2 is first that it is supported by empirical studies and second that in estimating the critical income level care has been taken to avoid accusation of exaggeration. However, in section 6.11, the robustness of results to changes in the value of e are examined.

B. <u>The critical consumption level (b):</u> The methods recommended by Little and Mirrlees have been described earlier. A version of these methods has been adopted by Scott, MacArthur and Newbery (1976) in a detailed way for Kenya. In Bangladesh however, the income levels at which direct consumption subsidies are offered are unclearly designated and income taxes are not applied to agriculture. The alternative approach used here is direct estimation of the critical income level (b) as a poverty line, from data on household expenditure and on the costs of a minimum diet.

As seen in Chapter Three, a World Bank estimate (IBRD 1980) of the poverty line suggests that 86.7 per cent of rural households would lie below it. As a tool for identifying the critical income level this is rather useless since the weights on consumption at any plausible value of e would be very large indeed and the resulting consumption benefits swamp other considerations. If a poverty line measures the lowest income at which people succeed in living then it is rather difficult to explain why most people in Bangladesh continue to live at measured income below this line. As Lipton (1983a) has shown in considerable detail the dietary minimum calorie intake of 2122 used by this World Bank study is an exaggerated estimate of the true per capita dietary minimum - and once allowance is made for actual body weights and activity patterns (not average ones) and for physiological adjustments to prolonged under-nutrition etc, the average requirement may be much lower especially for poor people. It is as well to point out at this stage that overestimates of the number of people in poverty are also because the data (household expenditure surveys) frequently fail to include all sources of income, especially some imputed sources from livestock, home gardens and casual labour, and sometimes fails to value them sufficiently highly.

As it happens, the World Bank study includes a second set of numbers based on a poverty line diet of 1805 calories (their hard-core poor at 85% or below of their dietary norm). Following Lipton (1983a) it is recognized here that this too is probably too large as an indicator of those at nutritional risk and his recommended method is followed of identifying the ultra-poor as those consuming less than 80 per cent of this sort of level. Expenditure equal to 80 per cent of 1805 calories per day per adult male equivalent1/ is thus the critical income level or in Lipton's terminology, the Food Adequacy Standard2/. More specifically it is accepted that these ultra-poor households are, by the criteria used, safely identifiable as those where few people would argue that their consumption, at the margin, is at least as valuable as uncommitted social income. As defined, the method adopted here is less egalitarian than the Little and Mirrlees approach cited earlier; however, by a different route, it is essentially aimed at measuring the same thing, the income level at which extra consumption is equally as valuable as uncommitted social income.

At 1976-7 prices this critical income level is Tk 1150[3]/ per head per annum (the IBRD study is carefully prepared and per capita estimates are based on detailed expenditure data and weighted by age, sex and activity group characteristics of rural households). There has been a 46 per cent increase in the consumer price index[4]/ between 1976-77 and 1979-80 which gives a critical consumption level in average mid 1979-mid 1980 prices of Tk 1679.

From the project survey on female wage labour earnings, (see Chapter Seven), average annual household earnings in the year to August 1980 were Tk 3448 (of which only TK170 on average were crop earnings). Average family size was 3.95 representing an average of 3.25 consumption units with an average conversion factor to consumption units of 0.823; (these families were smaller than the average size of 5.93 and had fewer dependents than the average family with 4.55 consumption units or 0.78 consumption units per family member).[5]/

The average earnings per consumption unit were therefore Tk 1061 which is just 63 percent of the critical consumption level. It should be noted that no allowance has been made for any unusual non-food contingent expenditure yet the income is still far from sufficient to meet the very low critical consumption level; on the other hand, any inadequacies in the survey data collected on a weekly recall basis, are likely to be in the form of omissions of income. If e =2 the weight for marginal additions to or losses from the incomes of these households is $(\frac{1679}{1061})^2$ which equals 2.50.

One unresolved question of method here regards the application of cost of living indices to workers whose payments are primarily in kind. If physical quantities of wages remain constant then the index should not apply to those elements. However, during field work female wage labour frequently asserted that the physical quantities they received per hour of work had also been falling over time. Clearly, there is a limit beyond which they cannot fall but when there are complex patron-client type relationships between the women who hire and the women who are hired the very notion of wage rates becomes suspect. However, even if, for this reason or because non-food expenditure (25%) has been overestimated, the actual food value of wages is ten per cent higher, the value of W_{fw} is still greater than 2.

To provide a similar estimate for mill workers and mill owners requires data on their family size, consumption units and household earnings. This data is not available. The weight (W_{mw}) adopted of 0.75 for mill workers is the average weight for rural households in Bangladesh (see Section D below and appendix footnote ten); it implies (at e=2) an income per consumption unit of approximately Tk 1939; mill workers annual mill earnings (Table 6.7) were Tk 3981 and therefore there is an implicit assumption of 2.05 consumption units per adult male equivalent wage earner. (It is unlikely that these households benefit from any female wage labour earnings and if they have the average family size in Comilla of 5.93, equivalent to 4.55 consumption units, then this dependency ratio requires 2.2 ($\frac{4.55}{2.05}$) workers at this wage rate per family. This is a high ratio but even quite large adjustments to this guestimate weight will not affect conclusions.)

In the case of mill owners there is a more clear-cut shorter and simpler argument for their weight. By any standard for rural Bangladesh mill owners are relatively rich and there is no justification for attaching a high social value to their incremental consumption. Therefore the weight used for mill owners is the weight for the highest expenditure class in the survey results referred in IBRD (1980, p.197); this weight, as given in the table in appendix footnote ten, is 0.11.

C. The savings premium (v): v measures the value of uncommitted social income in terms of average consumption. It is the ratio of the current value of the consumption stream generated over time from a unit of investment to the value of the same unit devoted to current consumption at the average level. However, once the critical consumption level has been specified, a value of v is implied for any current consumption level. Recalling that the critical consumption level, measuring the income level at which an incremental unit of consumption is equally as valuable as a unit of uncommitted social income, in this case is Tk 1679, then 1/v equals $(\frac{1679}{1061})^2$ for hired female labourers and v is 0.4. Applying the same formula to the (implied) average income per consumption unit of hired mill workers yields a value for v of 1.33. The reason for this difference is of course because v is defined in terms of the average consumption level and it is only at this level that d=1.

In applying an independently calculated value for v to any specific project there is an implicit presumption that the net effects of the project on consumption are on average consumption levels. In the *dheki* and huller mill comparison this is clearly inappropriate since with the *dheki* the consumption of the main group affected is five times more valuable relative to social income than those employed in huller mills. In fact, using World Bank data (IBRD 1980, p.185), and the consumer price index, (GOPRB 1983a, p.356) the average consumption level per capita in 1979-80 was estimated to be Tk 1791 which implies a v of 1.14. $(\frac{1679}{1791})^2$ is equal to 0.88 which is the weight on consumption at the average level which, in turn, is equal to $\frac{d}{v}$. (Since d = 1 at the average consumption level, v is equal to the reciprocal of 0.88 which is 1.14). The relevant question is whether this sort of value - close to unity - is a reasonable estimate of the savings premium in Bangladesh. If it is not then the critical consumption level (b) is imprecise; if v is considered to be too low it implies that b is too high. In principle, the value of v can be independently calculated according to the formula.6/ In practice, the formula will yield a value for v of infinity for Bangladesh since the rate of growth of per capita consumption is negative7/. Alternative approaches to measuring v using planned consumption growth and using micro-data were also unsatisfactory.8/

Another way of looking at v is in terms of the public investment constraint. If there are many socially profitable projects which are not being undertaken because of a shortage of domestic savings a low value for v is unlikely. In the Bangladesh case however domestic savings are supplemented by aid flows which are very very large in relation to public investment-in recent years actual aid flows have sometimes been larger than total development expenditure implying the government is also meeting establishment costs out of this aid and even so, there is a long "aid-pipeline" i.e. disbursements lag behind commitments. This problem has been shown to be more severe in Bangladesh than any other Asian developing country (IBRD 1984b).

The constraint on faster implementation of development projects is not the shortage of investible funds but the capacity of the administration to effectively use the funds that are available. Under these circumstances some studies (see Little and Mirrlees 1974, p.250) have used a value for v of 1. In the light of these considerations, and noting that the estimated critical consumption level is by no means generous, we conclude that the v implied here of 1.14 is a reasonable estimate.

D. The accounting price for private investment (APPI): An APPI has to be estimated for the *dheki* investment and for the huller mill investment. Since the people making these different investments come from different consumption groups it is to be expected that these two APPIs will be different. Apart from the application in converting fixed investment costs from efficiency prices to social prices the APPI for huller mill owners ($APPI_o$) is also used a second time to express the social value of their private savings9/ from the additional income they generate by investing in the huller mill. As noted above, because the households that hire female wage labour are a small and relatively rich subset of the households that invest in _dhekis_ two values need to be calculated for _dheki_ owners. W_{fh} is the weight given to employees of female wage labour and W_{di} is the weight given to all _dheki_ investors. The APPI is defined as the value of a unit of social income committed to private investment. To estimate it requires calculation of the social value of the most probable alternative use of the investment funds. In the case of the _dheki_ the investment has been seen to consist of 72 separate investments at a market cost of Tk 126 each. As suggested earlier, it seems reasonable to presume that generally the alternative use of these small sums would have been for consumption, rather than some other investment; hence this APPI is equal to W_{di}, the value in terms of social income of a unit of additional consumption by _dheki_ investors.

To estimate this value the distribution of _dheki_ investors by income level is required. In the first instance this distribution was taken to correspond approximately to the distribution of rice farmers-and information was available on the distribution of rice farmers by decile groups of gross income per consumption unit. On inspection of the results it was observed that this distribution is extremely even, with only the lowest and highest two deciles having a share of rice farmers outside of the 9.5 to 10.5 per cent range. It was decided therefore that a simple weighting by household per capita income levels would be reasonably accurate. It also had two advantages. First, it avoided relating two separate sets of income distribution data that would have been necessary in comparing the distribution of rice farmers by decile group income averages from the project survey with the distribution of rural incomes from World Bank data (IBRD 1980) used to estimate the critical consumption level. Secondly, the World Bank figures on the cumulative distribution of per capita income are not given for decile group but in twelve uneven size-class intervals; using decile groups would necessarily involve some inaccuracies, and the consumption weights are an exponential function of relative income so the inaccuracies could be serious. Once the population distribution is accepted as fairly reflecting the distribution of dheki investors then World Bank data on income levels and their size class intervals can be used. This is the approach adopted. The differences between the income distributions of dheki investors, and rural households will be the cause of a slight upward bias in the estimate, since the World Bank data is based upon a cumulative population distribution not household distribution-and poorer households are larger. An (arbitrary) ten per cent reduction in the value of W_{di} is therefore applied to the results.

The cumulative population distribution of per capita rural expenditure is given for 1976-77 (in 1973-74 prices) in IBRD (1980 p197). These values are adjusted from a per capita to a per consumption unit basis (by dividing by 0.78 - see Section B above and appendix footnote five) and multiplied by the increase in the cost of living index (2.07) to allow comparison with the critical consumption level in 1979-80 prices of Tk 1679 per consumption unit per year. The IBRD data is given for the average per capita income of twelve class intervals. The weight for each of these class means (x) is estimated separately according to $(\frac{1679}{x})^2$. The average of these weights, weighted in proportion to the percentage of the population to which they relate, is 0.83.10/ Reducing this by ten per cent to allow for the bias upwards gives a value for W_{di} of 0.75.

This is a convenient point to provide an estimate for the weight (W_{fh}) attached to the consumption of households that hire dheki labour. W_{fh} depends upon the place in the income distribution of those households that hire female wage labourers. Only fourteen per cent of all dheki owning households hire female labour and they will typically come from amongst the richest rural households - using the same data as for W_{di} the value W_{fh} for the fourteen per cent richest rural households, is estimated to be 0.17 (see appendix footnote ten).

In the case of the $APPI_O$ a slightly different approach has to be adopted. There is no information available about mill owners income levels but they are presumed to be rich and in section B a weight of 0.11 was put upon their additional consumption. However, the alternative use of the huller mill investment funds, Tk 7787 at efficiency prices, cannot be automatically taken as equal to this consumption weight as it is possible that non-consumption uses were an important element of alternatives to huller mill investment.

Little and Mirrlees (1974, pp.195-202) provide a detailed procedure for estimation of the APPI and the essence of the approach is to take a weighted average of the social values of the main alternative uses of the investment funds. The alternative uses of huller investment funds are taken to be equivalent to the sources of these funds; these sources are given in Shahnoor (1980) for the sample mills and their percentage shares are Business Profit (42%), Sale of Land (32%), (formal sector) Loans (13%), Sale of Gold (6.5%) and Personal Savings (6.5%). Loans and Personal Savings are funds that, through the banking system, would have been available for public sector investment - a main borrower from the banking system - and have a weight of one. The social value of funds raised through sale of land and gold requires knowledge of who the purchaser of land and gold is and what he would otherwise have done with his funds. This information can only be guessed but, fortunately, the results are not sensitive to poor guesswork since this weight only plays a small role in the calculations; on the assumption that he would have bought the same items from somebody else and that he is as rich as the mill owner these non-productive uses are given the consumption weight of 0.11. The social value of savings out of business profits (trading generally) is itself dependent upon the $APPI_o$ and is derived simultaneously. Given these values the $APPI_o$ is estimated to be 0.41[11]. (Properly, the $APPI_o$ is estimated at efficiency not market prices; the only adjustment required here, however, is to apply the consumption conversion factor (1.08) to those savings (38.5%) used for consumption. This adjustment is too small to alter the value of $APPI_o$.)

Appendix Notes

1. In fact this is given in the World Bank study as a per capita figure after adjustment for age, sex and activity group composition of rural households. Since these calculations are based on norms set in 1972 which are much too high (Lipton, 1983a, pp.9-25) a further (somewhat arbitrary) adjustment is to treat this dietary norm as that for an adult male and adjust total household requirements by the ratio of consumption units to family size.

2. Lipton in fact proposes a Food Adequacy Standard (FAS) in which the ultra-poor are defined as those who, spending 80 percent of their income on food, are consuming less than 80 per cent of the dietary norm. The data, from Household Expenditure Surveys, used by IBRD (1980 Annex 2 Table 7) is only broken down into four expenditure groups none of which spend 80 per cent of their incomes on food. It is noteworthy however that in 1976-77 the lowest forty per cent spent a lower percentage on food (76.5%) than the next forty per cent (78.8%). (This supports Lipton's (1983a p.2) identification of discontinuities at levels of ultra-poverty since from other IBRD data (1980 pp.174, 184 and 197) it can be estimated that in 1976-77 a similar percentage, in fact forty-five percent, of rural households had total purchasing power below the FAS, 80 per cent of the dietary minimum-and the number and percentage shows a rising trend. This is a high percentage and suggests that the Household Expenditure Survey may have underestimated outlays and probably failed to account adequately for subsistence production and kind payments. Over five per cent of households are estimated to be 45 per cent short of the FAS used here which is an impossibly low food intake level. On the other hand Bangladesh is extremely poor and her percentage of ultra-poor will be much higher than her neighbours.)

3. The IBRD-defined diet of 2,122 calories is not normative but actually based on household expenditure surveys and in 1973-74 prices cost Tk 2.45 per day, (IBRD 1980 p.173). According to their data for the bottom forty per cent of rural households, the cost of living increased by 42 per cent between then and 1976-77 the latest date for which their cost of living index is available (IBRD, p.184). The critical income line adopted is 80 per cent of their 85 per cent line or a diet of 0.8 x 1805=1444 calories per day. Together these adjustments yield a 1976-77 critical food consumption expenditure level of 2.45 x 365 x 0.85 x 0.8 x 1.42=863 takas. This makes no allowance for non-food expenditure which according to the World Bank study is about twenty-five per cent of all rural expenditure. The total income requirement is therefore Tk1150.

4. In fact IBRD (1980, pp.181-182) provides an improved set of weights to recalculate government indices used here (GOPRB, 1983a, p.356). However, the government indices have been adhered to since the data for the new weights for 1978-1980 is not available. For most years where comparisons were possible they gave very similar

results. This note applies to all future references to the cost of living index.

5. The data on average family size was from the preliminary survey of 4 villages (1638 households) in Comilla and, in the absence of meaningful estimates of "workers" and "non-workers" dependency ratios have to be inferred from the ratio of consumption units to family size. This ratio will get smaller as the number of dependents increases-since children have lower consumption units. Consumption units were estimated using the weights of 1 (adult males); 0.83 (adult females and children 10-14); 0.70 (children 5-9); 0.50 (children 1-4); 0.00 (children below one).

6. The formula for v is:

$$\frac{MPC\ (MP^{cap})}{CRI - MPS\ (MP^{cap})}$$

where MPC = Marginal propensity to consume.
MP^{cap} = Marginal productivity of capital.
CRI = Consumption rate of interest.
MPS = Marginal propensity to save.

Note: A) Except for the omission of the subtraction of unity this formula is that given by Hansen (1978 p.65) and, in this form, is the same as that used by UNIDO (1972) and Little and Mirrlees (1974 p.253) but has the advantage of using a more familiar notation. B) This formula is only valid if the CRI is greater than MPS (MP^{cap}). It also assumes that all values are constant through time (which is an impossible but not particularly harmful interpretation of true relative values). C) The CRI is defined as the product of the rate of growth of per capita consumption and the income elasticity of the marginal utility of income.

7. IBRD (1979b p.iv) estimates that per capita consumption has declined at the rate of 0.4 per cent per year during the last two decades. The shortened formula for v (footnote 6) is no longer valid [since the CRI will be negative whereas the formula requires the CRI to be greater than MPS (MP^{cap})]. The full formula (Little and Mirrlees, 1974 p.252) yields a value of infinity for v.

8. If the planned rate of increase in per capita consumption per annum of 2.24% from the five year plan is

used (GOPRB, 1980 pIII-9) the CRI given e=2 is 4.48. Estimating the marginal product of capital in the manner suggested by Squire and van der Tak (1975 pp.110-111) gives an MP^{cap} of 0.26, using data for 1960-1976 from IBRD (1979b p.44). With a consumption conversion factor of 1.08 these figures yield a lower estimate (using the equation for v, ignoring reinvestment, given in Squire and van der Tak 1975 p.106) of 5.4 for v. However, Squire and van der Tak say their method probably gives an overestimate of MP^{cap} and it is extremely optimistic to presume that planned increases in per capita consumption will actually occur when recent history suggests that they will not. An alternative estimate of v can be prepared using micro-data as discussed in Little and Mirrlees (1974 pp.247-250) and Squire and van der Tak (1975); an attempt was made to do this but the data were very unsatisfactory and the value derived (v=3.2) was considered unreliable. Largely due to data problems none of these direct estimates were felt to be sufficiently reliable to justify giving them a probability weighting relative to the imputed value based on the critical consumption level; i.e. they were ignored.

9. Other than the payment to hired wage labour no further use of W_{di} is required since the other income flows affecting dheki investors are imputed. The efficiency prices of all family labour are imputed values with no actual income flows, and therefore do not require use of consumption weights - in terms of the shadow wage formula c is equal to m.

10.

Average Monthly per Capita Consumption 1976-77 (in 1973-74 prices)	Average Monthly Consumption Per Consumption Unit 1979-1980 prices	value of W	% Population
(1)	(2)	(3)	(4)
16	42	11.11	0.8
30	80	3.06	1.2
37	98	2.04	2.8
41	109	1.65	5.3
45	119	1.38	6.6
52	138	1.03	14.1
58	154	0.83	14.1
72	191	0.54	23.7
88	233	0.36	12.5
109	289	0.23	10.6
137	363	0.15	4.0
160	424	0.11	4.3

Notes: a. Columns one and four are based on IBRD 1980 p.197.

b. Column Two is Column One x 2.07 x 1.28 to adjust respectively to 1979-80 prices and to consumption units.

c. Column Three is $(140 \div \text{Col.2})^2$ since the monthly critical consumption level is Tk 140 and e=2.

d. Column three is weighted by the values in Column Four to give a weighted average W_{di} of 0.83. This is adjusted downwards by ten per cent (see section D) to give a value for W_{di} of 0.75.

e. The value for W_{fh} had to be interpolated since the size class intervals don't give an exact weight for the upper 14 percent. It was estimated from the top three size classes as (4.3 x 0.11) + (4.0 x 0.15) plus (5.7 x 0.23) divided by fourteen giving a value of 0.17.

11. The weights are:-

Business Profits	=	.42 x $APPI_o$
Sale of Land	=	.32 x 0.11
Loans	=	.13 x 1.00
Sale of Gold	=	.065 x 0.11
Personal Savings	=	.065 x 1.00

$APPI_o$ = (.42 x $APPI_o$) + (.385 x 0.11) + (.195 x 1) which yields an $APPI_o$ of 0.41.

7

Policy Implications: Programmes for Rural Women

7.1 INTRODUCTION

Postharvest policy dialogue is chiefly concerned with the issue of food loss prevention. This is true in developing countries generally (Chapter Two) and in Bangladesh specifically (Chapter Three). The evidence from Bangladesh (Chapter Four) shows that farm-level food losses are low and do not offer much scope for cost-effective food loss prevention programmes. New postharvest technology is used by farmers because it saves labour and reduces farm management constraints, even though food losses may increase - as when pedal threshers are introduced (Chapter Five). But, higher capital-output ratios and the associated increases in labour productivity with new techniques, which on economic efficiency criteria are desirable, also result in adverse distributional changes in consumption (Chapter Six); the net effect is that aggregate consumption, weighted by consumption groups, is reduced.

One of the underlying themes in the presentation of material in these chapters and dealt with in some detail in Chapter Two, has been the need to broaden postharvest policy dialogue and to incorporate issues other than food loss. The distributional implications of new farm techniques have for a long time been a focal point of policy dialogue in, for example, irrigation and farm power; the preoccupation with farm-level food loss has inhibited this focus in postharvest policy dialogue. The evidence of the last chapter provides compelling proof of the need to reformulate the conventional food-loss concern but considerable care is required in identifying the future policy implications of these results.

Chapter six establishes only that technical change currently occurring with the substitution of the <u>dheki</u> by the mill has adverse income distribution effects; it does not necessarily follow that, on welfare grounds, policies should be adopted to prevent this substitution taking place. One alternative policy would be to promote alternative sources of earnings for the income groups adversely affected.

The focus of this chapter is on the identification of alternatives policies to redress the displacement of female wage labour by huller mill technology. The approach adopted is informed by three considerations. First, direct intervention to curtail farmer use of huller rice mills is not possible. It is not possible because the regulations required are complex (traders still need their mills), and incapable of being policed effectively. Direct intervention is not desirable to the (farmer and miller) majority of the rural population who benefit from the huller and may not be desirable to a majority of rural planners who spend their days devising ways to increase rural labour productivity. Planners who find the value judgements of chapter six acceptable and that intervention to stop huller mills is desirable in principle will also be aware that the control areas (Sen, 1984, p.224-229) they command are constrained and policy implementation cannot be juggled into shape in the same way that a social benefit-cost analysis can. According to Sen, (1984, p.229) the presumption implicit in SBCA as a planning tool is that the planner is either very powerful (and can insist upon compliance) or very stupid (and without a policy-design function). The SBCA presented in chapter six was not of course part of any Bangladesh planning exercise but for Sen-type reasons, if for no other, planners would not interpret the results as requiring a huller-curtailment policy.

Secondly, whilst some scope exists for improving the way in which food loss issues are addressed (see Section 4.10), a much more fundamental change in overall postharvest research prioritisation is required; the solution to problems that originate in the postharvest sector may lie, as in the case described below, in rather different sectors. Different, most critically, in the sense that usually a different set of researchers, planners and implementors of policy are involved. The underlying issue here is the institutional structure in which research is organised, objectives are set and the research-policy dialogue is conducted. Generalized remarks on reorganizing research, other than very limited ones such as the need to incorporate income distribution analysis, are of little value but the link between postharvest research and research on women, a feature of this study, is sufficiently universal to suggest one direction for development.

Thirdly, and related, whilst the choice of techniques exercise has demonstrated a policy-intervention need, it has told us little about the specific design of such policy. This chapter attempts to place the issue of postharvest technical change and rural women in the broader context of a) the gender division of labour (Section 7.2); and b) the process of peasant differentiation now occurring in Bangladesh (Section 7.3).

The analysis is concerned to show that the loss of employment opportunities because of the mills is but one component of a series of interacting changes occurring that are deepening the poverty of some rural households; it is in this context that the implications for women's programmes are discussed (Section 7.4).

7.2 THE GENDER DIVISION OF LABOUR

Through the arrangement of marriages, the laws of inheritance, in social customs and in decision-making, the social and household relations between men and women involve control of women's productive and reproductive activities; their (women's) passive support has been contrived through age hierarchies and cultural and religious heritage. Following Islamic precept, the control of female sexuality - organised through devices such as female seclusion (purdah), the kinship system and the gender division of labour - entails severe restrictions on women's physical mobility. The observance of these controls by women is an important determinant of the social status of her family and therefore there is enormous social pressure to conform; in effect requiring women to obey the wishes of the family patriarch.

The strength of patriarchal controls over women is most clear in the gender division of labour on the family farm. Traditionally, almost all female work is performed within the compound of the extended family and the major productive tasks they are responsible for include family and household care and maintenance, postharvest operations after threshing and livestock management. A women's duties in these areas are determined by her status within the household and in large part depend upon her position in the life cycle - daughter, wife, mother, mother-in-law, widow. The division of labour effort between tasks also depends upon family structure as well as farm size and seasonality.

More detailed descriptions of rural women's work are provided in Abdullah (1974), Kabir et al (1977), Khatun and Rani (1977), Alamgir (1977) and Abdullah and Zeidenstein (1982). Four characteristics of the Bangladesh gender division of labour are especially relevant for this analysis. First, the traditional division is extremely highly marked with very little substitution between occupations by gender. Typically, there is absolutely no involvement of women in field activities or in market activities and practically no involvement of men in activities around the farm yard or home.

This is most clearly seen in the most important agricultural operation, the production of rice. All activities from sowing up to the time of threshing are exclusively the responsibilities of men. Threshing marks the divide where both men and women may play a part but from threshing onwards all the operations through winnowing, parboiling, drying, husking, polishing, storage, cleaning and cooking are the responsibilities of women. However, where paddy is sold rather than consumed or when it is taken to the mill for husking and polishing rather than husked and polished at home, the responsibility is taken by the men of the household.

In other words, the division of labour in rice production is determined by the location of activity; all farmyard activity except threshing being the exclusive province of women. In other crops there are some exceptions to this. Women sometimes work with their husbands in the fields harvesting mustard seeds or at ponds away from the farmyard cleaning jute. But, these are rare exceptions where labour-intensive location-specific tasks have required poor families to use the labour of both sexes.

Secondly, unlike India where discrimination against female labour is primarily in access to work rather than in the wage (Lipton 1983b, pp.69-70) the rates of reward in Bangladesh for female wage labour are always lower than for men for similar time inputs. For example, a woman hired to husk rice for a 10 hour day will receive wages equivalent in value to less than three-quarters a man's wage for a day's labour (of eight hours at most) in weeding or cutting paddy. Traditionally, such female wage labour work operated as a social support system for widows and divorced women in households without a male farmer and therefore without crops of their own to process. For women the traditional workplace is their own farmyard and considerable loss of status is involved in working for others; conversely, high status is associated with freedom from physical work and employment of other women to undertake the farm work.

This attitude towards physical farm work is derived from a third aspect of the gender division of labour, namely the high status associated with female seclusion (purdah). The avoidance of farm work including own farm work makes it easier for women to avoid appearing in the presence of men other than those of the immediate family. This is a status consideration central to the social standing of the household. McCarthy (1978) has described the operation of purdah in rural Bangladesh and how changes are occurring in response to the changing economic status of households. For the poorest households the change agent is their poverty: their capacity to observe the restrictions on female mobility is constrained by the higher imperative of generating sufficient household income for daily food. The strict division of labour nevertheless restricts their wage labour to traditionally female tasks - particularly rice processing. The ideal of female seclusion remains and if the economic condition of the household improves the withdrawal of its women from wage labour invariably occurs.

The fourth feature of the gender division of labour is the continued low productivity of female work. Farm-level investment in improved seeds, chemical fertilizer and irrigation increase both the total demand for and productivity of male labour. There are three possible effects of increased agricultural output on women's work, none of which improve their productivity or their control over the distribution of farm income. The first is an increase in the volume of output they process using traditional technology without any improvement in labour productivity. The second is a switch to the use of male owned and operated commercial mechanized husking and polishing units in local markets; this reduces the workload of women, but as described above (Section 6.2) tends to displace wage labour rather than family farm women. The third situation is where the unprocessed rice (paddy) is sold and there is no increase in female work. The proportion of marketed surplus invariably increases with yield increases and is usually sold in the form of paddy which is then processed commercially. Thus, the only change in the gender division of labour occurs when husking and polishing is mechanized as the mills are always operated by men. The improvements in farm technology, whilst sometimes reducing the workload of women, make no contribution to improving their labour productivity or to the status of women as producers within the farm household.

To summarise, women's non-household productive labour on the family farm is rigidly determined by a gender division of labour in production of a joint-product requiring male labour in field operations; the absence of males or of land severely restricts their productive activities. Considerable loss of status is associated with deviations away from these gender-specific norms by seeking wage labour, which is anyway restricted to traditionally female activities. This gender division of labour facilitates the combination of family maintenance and farm activities by women and strengthens patriarchal control over women's mobility. Their contribution to farm output is under-reported, of low productivity and largely unaffected by modernising technical change.

7.3 PEASANT DIFFERENTIATION AND ITS EFFECT ON FEMALE WORK

This gender division of labour on the family farm was static for many years but the processes of rural change in Bangladesh are constraining its effectiveness as a basis for economic organisation. The most critical of these processes is population growth which has severely reduced the per capita cultivated area. According to the 1981 census (GOPRB 1981) the population was 87 million with a density of 1566 per square mile and an annual growth rate of 2.4 per cent. This is more than twice the population of 1951 and gives Bangladesh the highest population density in the world (apart from small island states). The average area of cultivable land per rural household is approximately 1.5 acres.

This land scarcity is being compounded by increasing inequality in land distribution. As described in Chapter Three, 30 per cent of rural households in Bangladesh own no cropped land and more than 50 per cent altogether own no cropped land or less than half an acre, (Jannuzi and Peach 1980). In effect, half the rural population is highly restricted in access to the means of production - land. Precise estimates of the rate of increase of landlessness are difficult to calculate because the definitions used in the various censuses vary and do not allow easy comparisons. However, one estimate (Begum and Greeley, 1979) shows that the number of households owning less than one acre has increased by over 230 per cent during the last thirty years and two other independent studies (Adnan et al, 1979,

Jannuzi and Peach, 1980) argue forcefully that the growth of landlessness is accelerating. Whilst tenancy agreements cover 24 per cent of cultivated land, the majority (over 80%) of tenant households are also land owning households; only 7.4 per cent of all farm households are pure tenants and they constitute less than one-fifth of landless households (Jannuzi and Peach, 1980).

Rural landlessness does not always involve poverty but nutritional (Ahmad 1977) and income (Khan 1977) measures of the growth of poverty have demonstrated that the incidence of household poverty has greatly increased with the growth of population and the increase in concentration of land ownership. The power of local patrons, usually large farmers, to control access to modern farm inputs, to control the availability of credit and to evade legal limits on land ceilings has prevented any realistic hopes for effective land reform. At the same time, the intensification of market relations of production is eroding traditional patron-client relationships in the rights over land and the use of labour.

In the absence of land, most households are entirely dependent upon the wage labour market. Their remaining assets are constantly at risk, for even short periods of unemployment cause distress sale of assets. Reduced labour earnings, for example, because flood- or drought-affected crops result in fewer wage labour opportunities in weeding and harvesting, cause households to meet consumption needs by sale of assets such as bullocks and boats which at other times would have increased their labour earnings. The seasonality of agricultural employment opportunities creates a cyclical pattern of debt which contingent consumption needs deepen; eventually this pattern erodes their resource base and leaves households without productive assets and dependent entirely upon their wage labour earnings.

This process of immiseration has been described in greater detail by Khan, (1977) and Alamgir, (1978). Specific concern here is with its effects upon the gender division of labour and, in particular, upon women's opportunities for productive work. The process of economic polarisation has not of course resulted in just two groups but in a continuum; at either end of this continuum one can identify two stylised types of the gender division of labour. The first, that of the farm household with land, maintains the traditional division of labour, but with increasing use of wage labour and with male family labour acquiring formal education and entering the urban labour market. The second group, without land, has a quite different set of patriarchal relations however.

Landlessness removes the opportunities for productive work on the family farm. Male labour then seeks wage employment. At the same time, the increasing competition for employment bids down wage rates. The only published estimates (Khan, 1977 and IBRD, 1980) on real agricultural wages (official statistics only record nominal wage rates) shows that they have been falling constantly since 1964. When family subsistence needs were being met at the minimum level and wage rates fall, an increase in labour effort is required in order to maintain family consumption at that minimum level. In the absence of family farm opportunities and with male income earning opportunities insufficient for subsistence needs this requires women's involvement in wage labour work. As described above, women's wage labour was traditionally restricted to women who were unable to obtain a subsistence share of farm production because of widowhood or divorce. This is no longer the case; in the project survey of eight villages in the Comilla and Tangail districts over eight per cent of the households depended upon female wage labour and in 63 per cent of those households the women who worked as wage labourers were still married. 69 per cent of these were also landless. In 62 per cent of the cases the increasing poverty associated with low male wage labour incomes was the stated reason why women worked as wage labour and ignored the status value associated with female seclusion. Some evidence of the nature and extent of this process is contained in Table 7.1. It is based upon a year-long survey undertaken in the eight project villages (see Begum and Greeley 1980a) of 100 households' earnings patterns and composed roughly equally of households where women did (WE households) and did not undertake (NWE households) wage work. The NWE households were also selected from amongst the poorest to provide some points of comparison that would help explain under what circumstances the process of immiseration forced female participation in wage labour.

Neither category had significant crop production. 18 WE households and 19 NWE households owned land but in both cases the average size of holdings was below 0.3 of an acre. The value of total crop production, including from leased in land, averaged less than 18 per cent of household earnings for those households growing crops.

Total earnings of the two categories were similarly low - only 6.3 per cent of the poverty line (see Appendix 6.1) but had very different gender compositions; in WE households only 47 per cent of household income was earned by men compared to over 82 per cent in NWE households and both the average number of males and average male earnings were very much lower in WE households. The low level of male earnings in WE households required female wage labour participation to meet minimum family subsistence needs and over the year nearly 25 per cent of non-crop income in WE households was earned by women; female children added a further 9 per cent. NWE households, which are equally poor, avoid this dependence primarily because family structure and the ownership of productivity-enhancing assets (boats, rickshaws, and small amounts of capital for petty trade or to buy raw materials for traditional village products) provide sufficient male income-earning opportunities. However, the absence of sufficient land to employ female labour productively in crop processing restricts their current use of female labour; any distress sale of assets that reduces male earnings will force them to drop their precarious maintenance of female seclusion and seek female wage labour earnings.

The aggregate data (Table 7.1) demonstrate the growing dependence of landless households on female wage labour. If women's earnings opportunities were plentiful, this dependence could ensure the economic security of the household though it would involve a substantial shift in the gender division of labour and in the form of patriarchy in so far as female seclusion was prevented. However, the rigidity in defining gender-specific tasks and the displacement of women's work by mechanisation of rice milling curtail these opportunities and therefore the capacity of female wage labour to prevent the pauperisation of landless households.

The regional differences provide some illustration of these constraints. In Chandina, Comilla where there are very few large farmers, the opportunities to obtain rice-processing work on other family farms were minimal and most of the earnings came from other sources, including one third of income from gleaning, begging and charity. In Madhupur, Tangail, the distribution of land is more concentrated and there were large farmers who employed female wage labour in traditional rice processing work; this accounted for 50 per cent of female wage labour earnings there. This difference between Chandina and Madhupur with respect to sources of female earnings also reflects differences in the development of the infrastructure and therefore of the intrusion of the market. In Chandina the more widespread distribution of (male operated) rice mills

Table 7.1
Household Earnings and the Composition of Female Earnings

A. Distribution of Household Earnings

	Annual Non-Crop		Percentage Distribution of Household Earnings			
	Household earningsb/ (Takas)c/	Nos of Adult Males per Household	Male Adults	Female Adults	Male Children	Female Children
WEa/ Households	3278	0.86	46.97	24.49	19.36	9.17
NWEa/ Households	3032	1.21	82.05	1.06	15.10	1.78

B. Composition of Female Earnings (WEa/ Households)

Source of Female Earnings	Per Cent of Total Female Earnings		
	Chandina	Madhupur	Both
Pure Rice Processing	10.2	37.9	28.9

Rice Processing and other crop processing combined	6.3	12.6	10.6
Jute, Millet and Mustard Processing	13.2	4.9	7.7
Household Work	17.4	29.9	25.8
Cleaning	19.7	0	6.2
Begging/Charity	12.9	6.1	8.3
Handicraft and other home-based earning activities	9.7	8.2	8.7
Sale of Fruit, Chickens, Eggs and Milk	10.6	0.3	3.5

Notes: a/ WE Households are those where women are wage earners. NWE households are those where they are not. In fact, as the Table shows, those women in NWE households earn a small amount through sale of fruit and dairy products.
b/ Sample sizes were 43 (WE households) and 57 (NWE households).
c/ There were approximately 30 takas to 1 pound sterling at the time of the survey.

Source: The IDS postharvest project survey during 1979/80 in Chandina thana, Comilla district and Madhupur thana, Tangail district.

allowed farmers to substitute the mill for hired female labour which restricted female employment opportunities and bid wages down. (As described in Section 6.2, during project fieldwork Madhupur received electrification and similar developments began immediately; in the next eighteen months the main market town of the region increased its number of rice mills from 4 to 20. At the end of this period one-third of farmers were using rice mills and of these nearly three quarters had previously hired female wage labour. More significantly, these farmers represented two-thirds of all those farmers that ever hired female labour.)

Paradoxically, just when the demand for female wage labour opportunities is increasing, the supply of such opportunities is being decreased. The two main sources of earnings for women in Chandina were gleaning (19.7%) and household work (17.4%) and no other single source of earnings accounted for more than ten per cent of their income. In Madhupur the dominant position of rice processing in female earnings is being eroded following the introduction of rice mills. Women there will face a similar bind to that now operating in Chandina imposed by the rigidity of the gender division of labour and the displacement of their labour through mechanization. These conditions severely restrict female earnings. In no month during the survey did the average proportion of days worked rise above 51 per cent and in the slackest month it fell to only 21 per cent. In 55.6% of the days without work the reason for not working was the non-availability of work.

This position of women in the wage labour market creates differences in their patriarchal relations in comparison to those women who have family farm work in addition to their domestic work. Landlessness prevents women from maintaining their traditional contribution to household income through the processing of crops. Poverty requires them to seek alternatives. Their opportunities to do so are restricted by the gender division of labour which strictly demarcates the earnings opportunities open to them. Limited access to non-land productive assets limits their home-based earnings and their dependence on crop processing and domestic work on larger farms results in low and uncertain wage incomes. Provision of base subsistence needs is often dependent on gleaning, begging and charity. This combination of landlessness, poverty and dependence of household income on low paid, low status and difficult to obtain female income reduces the stability of the patriarchal family in two ways.

First, the absence of land obviates the need for the patriarch of the family to have control over female labour for crop processing. Secondly, landlessness, poverty and dependence on female wage labour all contribute to an increase in the costs of labour reproduction. On family farms children can make a contribution to household earnings from an early age but in landless households they remain net consumers for longer. Absolute poverty does not diminish, and quite possibly increases, the ultimate value of children as income earners and as a source of security in old age; however, it reduces the resources available to raise children, causing high levels of infant mortality.

The dependence of the household on female wage earnings raises a conflict between reproductive and domestic responsibilities and the earning of wages away from the house. The fungibility of women's labour time between their different tasks, which is an essential element of the traditional gender division of labour, is weakened by the need to move out of the home to generate income. Whilst this increase in the costs of labour reproduction is mainly borne by women in the form of a longer working day, time allocation studies (Begum, 1983 chapter III) show that wage labour women with children spend fewer hours in income earning activities than women without children and fewer hours in domestic work than women who do not enter the wage labour market.

These changed economic relations within the family have provoked demographic shifts. Cain (1978) has shown that men from landless households are likely to marry later and die earlier than men with access to land; his study also confirms the higher infant mortality rate amongst the landless. Miranda (1980), in a careful analysis of recent demographic data, shows that there is a higher marriage instability amongst the poorest households reflected in the higher incidence of secondary celibacy below the age of thirty. According to the Bangladesh Fertility Survey (GOPRB, 1978c) marriage dissolution affected twenty-four per cent of women amongst the landless. Whilst marriage remains critical to the status and economic security of women, landlessness and the gender division of labour, by depriving them of productive employment, reduce their capacity to attain and retain a marriage contract.

7.4 IMPLICATIONS FOR WOMEN'S PROGRAMMES

One of the most contentious issues regarding the organisation of women's programmes in Bangladesh, as elsewhere, is the relative importance of economic class and gender class. The evidence of the last section suggests that female wage labourers have different problems, in reducing subordination, to the problems of women in landowning families. This is because women from landless households are deprived of their traditional form of economic contribution and have limited access to alternatives; also, the poverty of their households raises the relative costs of labour reproduction, increases the instability of the family and, therefore, the instability of the organisation of labour reproduction. In effect these women have little opportunity to conform to the ideals of patriarchal organisation of the family and the least to gain even if they do so.

The result is that landed and landless women often gain from different sets of policy initiative. The former may benefit most from programmes designed to reduce their subordination within the context of the family mode of production where there is extended fungibility (Lipton 1980) of assets between production and consumption. The latter require programmes that provide access to and control over the means of income self-reliance.

A possible rejoinder is the argument that reforming patriarchy, i.e. reducing female subordination, requires the active participation of women from the more prosperous landed classes precisely because it is their behaviour which determines the socially accepted norms of gender relations (Abdullah and Ziedenstein 1979, p.346). However, whilst it may remain the case that land ownership, a stable marriage and freedom from non-family work remain as ideals for these landless women, their realisation is an improbable prospect. The long term goals of programmes for women in Bangladesh most certainly concern changing the status associated with seclusion and women's exclusion from independent income-earning activities but it is the landless wage labour women who have most to gain and least to lose by participating in such programmes. Their attitude towards their status has already been forcibly changed by the imperative need to generate income.

Moreover, in an economically heterogeneous women's programme, the poor women would be exposed to the same class oppression in the distribution of resources as in male oriented programmes, e.g. of agricultural inputs. The development of women's programmes in Bangladesh has invariably been through a co-operative movement; when these have been economically heterogeneous it has been shown in practice (Feldman et al 1980) to mean that poor women get very few of the benefits from women's programmes.

From a policy perspective, there are two important points suggested by this analysis (Begum and Greeley 1980b). First, that wage labour women will invariably come from the poorest households and a focus on them as a category therefore provides a more effective means of targeting on poverty than other categories eg rural landless. Secondly, that these women will be more receptive than other women to programmes that require them to discard attributes and behaviour which conform to and perpetuate their subordinated status, because they have least to lose. In effect, programmes for landless women in Bangladesh provide a significant and unique opportunity to combine an attack upon female subordination and rural poverty.

Several local development agencies share this perspective on the role of programmes for wage labour women; their experience has been carefully reviewed (Begum 1981 and 1983, chapter IV) and a number of operating principles can be derived. Whilst these principles are most certainly not of universal application to the needs of wage labour women's programmes they are sufficiently well-tested, particularly by long established programmes such as those of the Bangladesh Rural Advancement Committee, to provide a useful point of departure. There are five points.

a) Insistence upon <u>single class</u> as well as single sex co-operatives. Programmes have achieved this by, for example, using physical work as a means to separate women according to whether status concerns or income generation are more important determinants of their occupational preferences.

b) Within the co-operatives the main emphasis has to be upon income generation in fairly immediate forms rather than through agriculture investment (when they have no land) or through human capital development (health and nutrition programmes).

c) Moreover, since they have no place in agricultural production on family farms and an increasingly reduced role as wage labourers, their activities have to be oriented towards the market rather than traditional forms of farm work.

d) They also have little risk bearing capacity and an essential part of programmes is some form of risk fund generated by savings whereby contingencies such as sickness or pregnancy do not result in sudden worsening of their economic condition.

e) Group rather than individual investments that will provide some protection from interference by class interests antagonistic to them. Group investment will also allow larger investments thereby more quickly increasing labour productivity.

None of these conditions would be particularly relevant to women from a higher economic class. In this, they still contain an essential paradox for the more successful they are the less relevant they become - but only if there has been no change in the status values determining women's position.

To address this problem, past programmes have included functional education directed towards consciousness raising and development of class cohesion. These have achieved some local success and in doing so made a contribution to more widespread changes in the structure of patriarchy. However, landless women's subordinated status is influenced by deeper seated social conditions. The most important of these, as this analysis has tried to argue, is the economic role of the family through the transition to full capitalist development. In a situation of labour surplus the growing immiseration of many families amongst the landless seems certain and differences in their reproductive behaviour compared to farm households are becoming clear. Women's close dependence upon the family and household assets for the value of their productive and reproductive activities is well established and the threat to them caused by the threat to the family is central to the crisis of Bangladesh's rural development.

Acronyms

AID	Agency for International Development
ASEAN	Association for South-East Asian Nations
BARC	Bangladesh Agricultural Research Council
BARD	Bangladesh Academy for Rural Development
BCSIR	Bangladesh Council of Scientific and Industrial Research
CIDA	Canadian International Development Agency
EEC	European Economic Community
FAO	Food and Agricultural Organisation
GOPRB	Government of the Peoples Republic of Bangladesh
GTZ	German Technical Aid Agency
IARI	Indian Agriculture Research Institute
IBRD	International Bank for Reconstruction and Development
IDA	International Development Association
IDRC	International Development Research Centre
IDS	Institute of Development Studies

IFPRI	International Food Policy Research Institute
IGSI	Indian Grain Storage Institute
IRRI	International Rice Research Institute
ODA	Overseas Development Administration
REB	Rural Electrification Board
SEARCA	Southeast Asian Regional Centre for Graduate Study and Research in Agriculture
SIDA	Swedish International Development Authority
TPI	Tropical Products Institute (now the Tropical Development Research Institute)
UNEP	United Nations Environment Programme
UNU	United Nations University

Glossary of Local Terms

Aman	Rice harvested in winter
Aus	Rice harvested in summer
Bari	Cluster of households with a common boundary
Boro	Rice harvested in early summer
Chakki	Wheat grinding stone
Dheki) Dholong) Khol Chiya)	Traditional rice husking methods
Haor	Large lakes (the edges of which are used for growing low-land rice)
Kula	Winnowing tray
Kutias) Barkiwalas)	People engaged in small-scale paddy-rice processing businesses.
Rabi crops	Non-rice winter crops
Raings	Aluminium or earthenware pots for parboiling paddy

References

Abdullah, Tahrunnessa Ahmed, 1974, "Village Women As I Saw Them". Ford Foundation, Dhaka.

Abdullah, Tahrunnessa and Sondra Zeidenstein, 1979, "Women's Reality: Critical Issues for Program Design." Studies in Family Planning, vol.10 no.11/12, pp.344-352.

Adams, J.M. and G.W. Harman, 1977, The Evaluation of Losses in Maize Stored in a Selection of Small Farms in Zambia with Particular Reference to the Development of Methodology. Tropical Products Institute, Report no.G109. London.

Addison, Joseph, n.d., A Quotation from an article in the Spectator no.477 cited in The Oxford Mini-Dictionary of Quotations. (1983), Oxford University Press, London, p.1.

Administrative Staff College, 1976, Postharvest Grain Losses. Supporting Study 12. All India grain storage and distribution study, Hyderabad, India.

Adnan, Shapan, H.Z. Rahman, Mahbub Ahmed and Malik Md. Shahnoor, 1979, "Review of Landlessness in Rural Bangladesh." Unpublished paper presented at the Conference on Basic Needs, Appropriate Technology and Agrarian Reforms, Dhaka.

Agency for International Development, 1977, "Expenditures in Reducing Postharvest Food Losses." Unpublished AID Report for Committee on Foreign Relations, United States Senate.

Agricultural Marketing Directorate, 1968, "Report of the Study Group on the Feasibility of Pilot Project concerned with Improved Methods of Harvesting, Drying, and Storage of Paddy and Rice at Farm-Level." Government of East Pakistan, reprinted by Bangladesh Agricultural Research Council, Dhaka.

Ahmad, Kamaluddin, ed., 1977, Nutrition Survey of Rural Bangladesh. Institute of Nutrition and Food Science, University of Dhaka.

Ahmad, Nafis, 1976, A New Economic Geography of Bangladesh. Vikas Publishing House pvt. Ltd, New Delhi.

Ahmed, Jasim U, 1981. "Postharvest Paddy Processing in Bangladesh: Is Mechanization Needed at Farm Level?" Bangladesh Journal of Agricultural Economics, vol. IV, no.I, pp.23-45.

_____________, 1982, "The Impact of New Paddy Post Harvest Technology on the Rural Poor in Bangladesh." Martin Greeley and Michael Howes, eds., Rural Technology,

Rural Institutions and the Rural Poorest, CIRDAP, Comilla, pp.105-127.

Ahmed, Jasim U., K. Hasan and G.W. Sarker, 1980, "Effects of Farm Size and Tenancy on Post-Production Loss as Estimated by Farmers: The Case of Aus Paddy." Paper presented at the Post-Production Workshop on Food Grains. Food Science & Technology Division, Bangladesh Council of Scientific and Industrial Research, Dhaka.

Ahmed, Raisuddin, 1979, Foodgrain Supply Distribution and Consumption Policies within a Dual Pricing Mechanism: A Case Study of Bangladesh. International Food Policy Research Institute, Research Report no.8.

Ahmed, Sadiq, 1976, "An Exercise in Social Profitability Analysis: The Case of Ashuganj Fertilizer Project." The Bangladesh Development Studies, vol.IV, no.4, pp.479-498.

Ahmed, Sufi Mohiuddin, 1977, "A Report on Survey of Wheat Seeds Stored by the Farmers for Sowing in 1976-77." Agricultural Research Institute, Dhaka.

Alam, Ferdous Md., 1981, "Cost of Credit from Institutional Sources in Bangladesh." Bangladesh Journal of Agricultural Economics, vol.IV, no.2, pp.51-62.

Alamgir, Mohiuddin, 1978, Bangladesh: A Case of Below Poverty Level Equilibrium Trap. Bangladesh Institute of Development Studies, Dhaka.

Alpert, M.E., M.S.R. Hutt, G.N. Wogan and C.S. Davidson, 1971, "Association Between Aflatoxin Content of Food and Hepatoma Frequency in Uganda." CANCER, vol.28, pp.253-260.

Andersen, P.C. and J. Hoberg, 1974, "Report on DANIDA Drying Project." BARD, Comilla.

Araullo, E.V., D.B. de Padua and Michael Graham, eds., 1976, Rice: Postharvest Technology. International Development Research Centre, Ottawa.

ASEAN, 1985, Food Handling News Letter, no.18 October.

Asia Foundation, The, 1980, "Energy from Rice Husks: 1980-83." A Project Proposal submitted to USAID, Dhaka.

Bala, Bilash Kanti, n.d., "Development of a High Temperature Rice Drier for Drying Raw Paddy as well as Parboiled Paddy." A research proposal submitted to the Bangladesh Agricultural Research Council from the Dept. of Farm Power and Machinery, Bangladesh Agricultural University, Mymensingh.

Bala, Bilash Kanti, M. Daulat Hossain and A.T.M. Ziauddin, 1980, "Design and Performance of a Deep Bed Drier for Paddy Using LP Gas Burner." Paper presented at the

Post-Production Workshop on Food Grains. Food Science & Technology Division, Bangladesh Council of Scientific and Industrial Research, Dhaka.

Bangladesh Academy for Rural Development, (n.d.), "Post Production Loss." Unpublished paper, Comilla.

Bangladesh Power Development Board, 1977, Rural Electrification Feasibility Study. vols. I and III. Commonwealth Associates Inc. NRECA International Ltd, Dhaka.

Becker, S. and M.A. Sardar, 1981, "Seasonal Patterns of Vital Events in Matlab Thana, Bangladesh." In Robert Chambers, Richard Longhurst and Arnold Pacey eds., Seasonal Dimensions to Rural Poverty. Frances Pinter, London, pp.149-153.

Begum, Saleha, 1981, "Report on a Study of Women's Organisation in Bangladesh." Swedish International Development authority, Dhaka.

____________, 1983, "Women and Rural Development in Bangladesh." Unpublished M.S. dissertation, Cornell University.

Begum, Saleha and Martin Greeley, 1979, "Rural Women and the Rural Labour Market in Bangladesh: An Empirical Analysis." Bangladesh Journal of Agricultural Economics, vol.11, no.2, pp.35-56.

____________, 1980a, "Women, Employment, and Agriculture: Notes from a Bangladesh Case Study." Unpublished paper presented at a Seminar organized by the Women for Women Research and Study Group, Dhaka.

____________, 1980b, "Co-operatives for Wage-Labour Women: Outline of a Proposal for IRDP." Institute of Development Studies, University of Sussex, UK. Unpublished Paper.

Bell, Clive, Peter Hazell and Roger Slade, 1982, Project Evaluation in Regional Perspective, A World Bank Research Publication, The Johns Hopkins University Press, London.

Bergeret, A., 1981, "Where are the Sahel Grain Losses Located." Industry and Environment, vol.4, no.1, pp.16-18.

Bhalla, S.K. and T.B. Basnyat, 1982, "Postharvest Technology and its Impact on the Rural Poor in Nepal." In Martin Greeley and Michael Howes, eds., Rural Technology, Rural Institutions and the Rural Poorest. CIRDAP, Comilla, pp.183-195.

Binswanger, Hans P., 1978, The Economics of Tractors in South Asia. Agricultural Development Council, New York and International Crops Research Institute for the Semi-arid Tropics, Hyderabad.

Birla Institute, 1979, Losses of Foodgrains in India. Birla Institute of Scientific Research, Economic Research Division, New Delhi.

Bose, Swadesh R, 1968, "Trend of Real Income of the Rural Poor in East Pakistan." The Pakistan Development Review, vol.VIII, no.3, pp.452-488.

Bourne, Malcolm C., 1977, Post Harvest Food Losses - the Neglected Dimension in Increasing the World Food Supply. Department of Food Science and Technology, Cornell University, Ithaca, New York.

Boxall, R.A., M. Greeley, D.S. Tyagi, with M. Lipton and J. Neelakanta, 1978, "The Prevention of Farm-Level Food Grain Storage Losses in India: A Social Cost-Benefit Analysis." IDS Research Report, University of Sussex.

Boxall, R.A., Martin Greeley and J. Neelakanta, 1976, "Report on Visit to Bangladesh by IDS/IGSI Grain Storage Team: Review of IDS/IGSI Storage Project and Observations on Farm-Level Storage in Bangladesh." Appropriate Agricultural Technology Cell, Bangladesh Agricultural Research Council, Dhaka, Monograph no.2.

Breese, M.H., 1960, "The Infestability of Stored Paddy by Sitophilus Sasakii (Tak.) and Rhizopertha Dominica (F)." Bulletin of Entomological Research, vol. 51, no.3, pp.599-630.

Brown, Lester, 1970, Seeds of Change. Pall Mall, London.

Burch, David, 1978, "Overseas Aid and the Transfer of Technology: A Case Study of Agricultural Mechanization in Sri Lanka." D.Phil. thesis, University of Sussex.

Cain, Mead T., 1978, "The Household Life Cycle and Economic Mobility in Rural Bangladesh." Centre for Policy Studies Working Papers, Population Council, New York.

Calpatura, R.B., 1981, "Variety, Maturity and Length of Straw Cutting Interaction in the Grain Losses of Rice during Harvesting." Proceedings of the 4th Annual Workshop on Grains Postharvest Technology, SEARCA, Laguna, Philippines, pp.373-382.

Calverly, D.J.B., P.R. Street, T.J. Cree and D.A.V. Dendy, 1977, "Postharvest losses of rice in Malaysia." Paper presented at the Conference on Food and Agriculture, Malaysia.

CERES, 1977, "World Report." CERES no.60, pp.4-14.

Chaudhury, R.H., 1981, "Seasonality of Prices and Wages in Bangladesh." Robert Chambers, Richard Longhurst and Arnold Pacey, eds. Seasonal Dimensions to Rural Poverty. Frances Pinter, London, pp.87-91

Chayanov, A., 1966, The Theory of Peasant Economy. Edited by Daniel Thorner, Basile Kerblay and R.E.F. Smith. American Economic Association, Translation Series, Irwin, Homewood.

Chowdhury, Amirul Islam, 1980, "Rural Development: Policies, Strategies, Institutions and Programmes." Qazi Kholiquzzaman Ahmed ed., Development Planning in Bangladesh - A Review of the Draft Second Five Year Plan. Bangladesh Unnayan Parishad, Dhaka, pp.26-37.

Chowdhury, A.K.A., S.L. Hoffman and L.C. Chen, 1981, "Agriculture and Nutrition in Matlab Thana, Bangladesh." Robert Chambers, Richard Longhurst and Arnold Pacey, Seasonal Dimensions to Rural Poverty. Frances Pinter, London, pp.52-60.

Chowdhury, Omar H., 1981, "Estimating Distributional Weights for Bangladesh." The Bangladesh Development Studies. vol.IX, no.2, pp.103-110.

Chung, C.J., 1980, Post-Production Rice Systems in Korea: Final report of Phase II. College of Agriculture, Seoul National University Suweon, Korea.

Clark, Stuart, C and Haridas Saha, 1980, "Solar Drying of Paddy" paper presented at the Post-Production Workshop on Food Grains. Food Science and Technology Division, Bangladesh Council of Scientific and Industrial Research, Dhaka.

Clay, Edward J. and W.A. Shah, 1976, "Wheat Storage and Methods of Storage in a Traditional and a New Wheat Cultivation Area of Bangladesh." Agricultural Development Council Inc., Dhaka.

Collier, William, L., 1979, "Choice of Technique in Rice Milling." Tan Bock Thiam and Shao-Er Ong, eds., Readings in Asian Farm Management, Singapore University Press, pp.333-345. Comment on Timmer, 1974.

Comptroller General's Report to the Congress (US) n.d., "Hungry Nations Need to Reduce Food Losses Caused by Storage, Spillage, and Spoilage." Department of State and other agencies, Washington D.C.

Coursey, D.G., 1981, "Traditional Postharvest Technology of Tropical Perishable Staples." Industry and Environment, vol.4, no.1, pp.10-14.

Dawlatana, Mamtaz, 1980, "Effect of Milling and Per Cent Brokens on Cooking Loss of Rice in Bangladesh." Paper presented at the Post-Production Workshop on Food Grains. Food Science & Technology Division, Bangladesh Council of Scientific and Industrial Research, Dhaka.

de Vylder, Stefan and Daniel Asplund, 1979, Contradictions and Distortions in a Rural Economy: The Case of Bangladesh. Policy Development and Evaluation Division, Swedish International Development Authority.

Dasgupta, Biplab, 1977, Agrarian Change and the New Technology in India, United Nations Research Institute for Social Development, Geneva.

Farouk, M.O., 1970, "Structure and Performance of the Rice Marketing System in East Pakistan." Occasional Paper, no.31, Cornell University, Ithaca.

Farouk, M.S., 1975, "Probable Losses in Postharvest Period: How to Minimise Them." Proceedings of the Workshop on Appropriate Agricultural Technology. Bangladesh Agricultural Research Council, Dhaka.

Farrington, John and Fredrick Abeyratne, 1982, Farm Power and Water Use in the Dry Zone. Part II. Agrarian Research and Training Institute, Colombo Research Study No.52.

Feldman, Shelley, Farida Akhtar, and Fazila Banu, 1980, The Research and Evaluation Study of the IRDP Pilot Project on Population and Rural Women's Cooperatives. CIDA, Dhaka.

Food and Agriculture Organisation, 1968, The State of Food and Agriculture. United Nations Food and Agriculture Organization, Rome.

__________, 1975, Reducing Postharvest Food Losses in Developing Countries, FAO, Rome.

__________, 1981a, Production Year Book, vol.34 (1980), FAO, Rome.

__________, 1981b, "Action Programme for the Prevention of Food Losses: Review of Loss Assessment Activities of PFL Projects in the Asia Region" Consultancy Report to the Agricultural Services Division, FAO by Z. Toquero.

Gaiser, D., 1981, "A Brief Summary of Paddy Loss in Indonesian Rice Postharvest Systems." Proceedings of the 4th Annual Workshop on Grains Postharvest Technology. SEARCA, Laguna, Philippines, pp.133-138.

Gariboldi, F., 1974, "Rice Parboiling." Agricultural Development Paper, no.97, FAO, Rome.

Giles. P.H., 1964, "The Storage of Cereals by Farmers in Northern Nigeria." Tropical Agriculture (Trinidad), vol. 41, no.3, pp.197-212.

Gill, K.S., 1982, "Postharvest Market Technology for Cereals (Paddy and Wheat) - Needed Improvement (The Punjab Case)." Martin Greeley and Michael Howes eds., Rural Technology, Rural Institutions and the Rural Poorest, CIRDAP, Comilla, pp.152-165.

Gittinger, J. Price, 1972, Economic Analysis of Agricultural Projects. The John Hopkins University Press, Baltimore and London.

Government of India, 1967, Interim Report of the Expert Committee on Storage Losses of Foodgrains During Postharvest Handling. Dept. of Food, Ministry of Agriculture and Irrigation, New Delhi.

__________, 1971, Final Report of the Expert Committee on Storage Losses of Foodgrains During Postharvest Handling. Dept. of Food, Ministry of Agriculture and Irrigation, New Delhi.

Government of the People's Republic of Bangladesh, 1972, Master Survey of Agriculture in Bangladesh. Bangladesh Bureau of Statistics, Dhaka.

__________, 1978a, "Summary Report of the 1978 Land Occupancy Survey of Rural Bangladesh." Bangladesh Bureau of Statistics, Dhaka.

__________, 1978b, Report of the Task Force on Rice Processing and By-Product Utilization in Bangladesh. Ministry of Agriculture, Dhaka.

__________, 1978c, Bangladesh Fertility Survey 1975: First Report. Ministry of Health and Population Control, Dhaka.

__________, 1979, Costs and Returns Survey for Bangladesh 1978-79 crops. Vols. I-V, Agro-Economic Research, Ministry of Agriculture and Forests, Dhaka.

__________, 1980a, The Second Five Year Plan 1980-85. Planning Commission, Dhaka.

__________, 1980b, The Year Book of Agricultural Statistics of Bangladesh 1979-80. Bangladesh Bureau of Statistics, Statistics Division, Ministry of Planning, Dhaka.

__________, 1980c, Bangladesh Agricultural Census, 1977: Preliminary Results. Bangladesh Bureau of Statistics, Dhaka.

__________, 1981, The Preliminary Report of Bangladesh Population Census, 1981. Bangladesh Bureau of Statistics, Dhaka.

__________, 1983a, Statistical Pocket Book of Bangladesh, 1982. Bangladesh Bureau of Statistics, Dhaka.

__________, 1983b, Tangail District Statistics. Bangladesh Bureau of Statistics, Dhaka.

Government of East Pakistan, 1965, Master Survey of Agriculture in East Pakistan, 1963-64 (second round). East Pakistan (now Bangladesh) Bureau of Statistics, Dhaka.

Greeley, Martin, 1978, "Appropriate Rural Technology: Recent Indian Experience with Farm-Level Foodgrain Storage Research." Food Policy, February, pp.39-49.

__________, 1981, "Farm-Level Rice Processing in Bangladesh: Food Losses, Technical Change and the Implications for Future Research." Proceedings of the 4th Annual Workshop on Grains Postharvest Technology. SEARCA, Laguna, Philippines. pp.33-47.

__________, 1982a, "Rural Technology, Rural Institutions and the Rural Poorest: The Case of Rice Processing in Bangladesh." Martin Greeley and Michael Howes, eds., Rural Technology, Rural Institutions and the Rural Poorest. CIRDAP, Comilla, pp.128-151.

__________, 1982b, "Farm-Level Postharvest Food Losses: "The Myth of the Soft Third Option." IDS Bulletin, vol. 13, no.3 pp.51-60.

__________, 1983a "Patriarchy and Poverty: A Bangladesh Case Study." South Asia Research, vol.3 no.1, pp.35-55.

__________, 1983b, "Solving Third World Food Problems: The Role of Postharvest Planning." Morris Leiberman, ed., Postharvest Physiology and Crop Preservation. Plenum Press, New York and London, pp.515-535.

__________, 1983c, "Report on the Socio-Economic Aspects of Implementation of the Bangladesh Farm- and Village-level Postharvest Rice Loss Assessment Project." First Consultancy Report for the FAO Prevention of Food Loss Programme, July.

__________, 1984, "Report on the Socio-Economic Aspects of Implementation of the Bangladesh Farm- and Village-level Postharvest Rice Loss Assessment Project." Final Consultancy Report for the FAO, Prevention of Food Loss Programme. November.

Greeley, Martin and Siddiqur Rahman, 1980, "Wet Season Postharvest Food Losses." Paper presented at the Post-Production Workshop on Food Grain. Food Science and Technology Division, Bangladesh Council of Scientific and Industrial Research, Dhaka.

Grist, D.H., 1975 Rice. Longman Group Limited, London.

Guggenheim, Hans, 1978, "Of Millet, Mice and Men: Traditional and Invisible Technology Solutions to Postharvest Losses in Mali." David Pimentel, ed., World Food, Pest Losses, and the Environment. Westview Press, Boulder, Colorado, pp.109-161.

Gupta, O.P., 1981, "Estimation of Food Grains Storage Losses." Indian Grain Storage Institute, Hapur, Uttar Pradesh, unpublished Paper.

Halpern, Peri, M., 1978, "Labour Displacement of Rural Women in Bangladesh: Possible Consequences and Possible Solutions - What Would Constitute An Effective Feminist Approach." Paper presented at Conference 133 on The Continuing Subordination of Women in the Development Process, Institute of Development Studies, University of Sussex.

Haque, A.K.M. Anwarul, M.A. Quasem, Nurul H. Choudhury and Jose R. Arboleda, 1985, "Rice Post-Harvest Practices and Loss Estimates in Bangladesh: Harvesting to Cleaning." Bangladesh Rice Research Institute and Food and Agriculture Organisation of the United Nations Postharvest Project, Field Document 6, Dhaka

Harris, Kenton L., and Carl J. Lindblad (compiled), 1978, Post-harvest Grain Loss Assessment Methods. American Association of Cereal Chemists, St. Paul, Minnesota.

Harriss, Barbara, 1974, "Operational Efficiency in Rice Milling Technologies in North Arcot District, Tamil Nadu." Paper from the Project on Agrarian change in Rice-Growing Areas of Tamil Nadu and Sri Lanka presented at a Seminar at St. Johns College, Cambridge. 9-16, December.

__________, 1976, "Paddy Processing in India and Sri Lanka: A Review of the Case for Technological Innovation." Tropical Science, vol.18, no.3, pp.161-186.

__________, 1978, "Rice Processing Projects in Bangladesh: An Appraisal of a Decade of Proposals." The Bangladesh Journal of Agricultural Economics, vol.1, no.2, pp.24-52.

__________, 1979, "Postharvest Rice Processing Systems in Rural Bangladesh: Technology, Economics and Employment." The Bangladesh Journal of Agricultural Economics, vol.2, no.1, pp.23-50.

Hazell, Peter, B.R., 1982, "Instability in Indian Foodgrain Production." IFPRI Research Report no.30, Washington, D.C.

Heady, Earl O., and John L. Dillon, 1961, Agricultural Production Functions. Iowa State University Press, Ames, Iowa.

Hossain, Mahabub, 1980, "Agriculture in the Draft Second Five Year Plan". In Qazi Kholiquzzaman Ahmad, ed., Development Planning in Bangladesh - Review of the Draft Second Five Year Plan, Bangladesh Unnayan Parrshad, Special Publication - I, pp.38-48.

Humphrey, Hubert, H., 1977, Letter to John Gilligan, Administrator, Agency for International Development, Washington D.C., October 12.

Huq, A.K. Fazlul, 1980, "Rice in Bangladesh: Estimation of Food Losses in Farm-Level Storage." Unpublished paper presented at the Post-Production Workshop on Food Grains. Food Science and Technology Division. Bangladesh Council of Scientific and Industrial Research, Dhaka.

Huq, A.K. Fazlul and Martin Greeley, 1980, "Rice in Bangladesh: An Empirical Analysis of Farm-Level Food Losses in Five Post-Production Operations" in Proceedings of SEARCA's 3rd Annual Workshop on Grains Postharvest Technology, Kuala Lumpur, Malaysia, pp.245-262.

Huq, M.S., and G.K. Joarder, 1979, "Evaluation and Estimation of Aflatoxins in Rice in Bangladesh." Food Science & Technology Division, Bangladesh Council of Scientific and Industrial Research, Dhaka. Unpublished paper.

Hurley, E.G., I.W. Wilcox, M. Abdul Quasem and M. Aminuzzaman, 1980, "Rice Postharvest Technology Project: Project Report for the Two Years 1978-79." Tropical Products Institute (UK) and Bangladesh Rice Research Institute, Joydebpur. Unpublished Paper.

Hurley, E.G., 1980, "Design of Paddy Driers." Paper presented at the Post-Production Workshop on Food Grains. Food Science & Technology Division, Bangladesh Council of Scientific and Industrial Research, Dhaka.

Indian Agricultural Research Institute, 1980, "Resistance to Storage Insects in Wheat Grain." Research Bulletin no.28, New Delhi.

International Bank for Reconstruction and Development, 1970, "Evaluation of Benefits from Season-to-Season Grain Storage Programs" Economics Dept. Working Paper no.66 Washington D.C.

__________, 1978, "Bangladesh: Report on Domestic Financial Resource Mobilization." Unpublished (grey cover) document of the World Bank, Washington D.C.

__________, 1979a "Bangladesh: Food Policy Issues." Unpublished (grey cover) document of the World Bank, Washington, D.C.

__________, 1979b, Bangladesh: Current Trends and Development Issues. John Hopkins University Press, Baltimore and London.

__________, 1980, "Bangladesh: Current Economic Position and Short-Term Outlook." Unpublished (grey cover) document of the World Bank, Washington, D.C.

__________, 1983, "Bangladesh: Selected Issues in Rural Employment." Unpublished (grey cover) Document of the World Bank, Washington D.C.

__________, 1984a, World Development Report 1984, Oxford University Press, London

__________, 1984b, "Bangladesh: Economic Trends and Development Administration." Unpublished (grey cover) document of the World Bank, Washington D.C.

International Rice Research Institute, 1978, A Report on Rice-Postproduction Technology Project. Submitted by the International Rice Research Institute and the University of the Philippines at Los Banos to the Bicol River Basin Development Program, Los Banos.

Irvin, George, 1978, Modern Cost-Benefit Methods: An Introduction to Financial, Economic and Social Appraisal of Development Projects. The Macmillan Press Ltd, London.

Islam M.N., 1980, "Study of Problems and Prospects of Biogas Technology as a Mechanism for Rural Development: Study in a pilot area of Bangladesh" mimeo., Bangladesh University of Engineering and Technology, Dhaka.

Islam, Rizwanul, 1977, "Shadow Wage Rates in the Manufacturing Industries of Bangladesh." Bangladesh Development Studies, vol. V, no.1, pp.51-80.

Jabber, M.A., 1980, "Research Report on a Field Survey of Rice Procurement Centres" Bangladesh Agricultural University, Mymensingh.

Jabber, Md. Abdul, 1982, "Rice Processing in Bangladesh Rice Research Institute Pilot Project Area, Joydebpur, Bangladesh." Unpublished M.S. dissertation, University of the Philippines.

Jannuzi, F. Tomasson and James T. Peach, 1980, The Agrarian Structure of Bangladesh: An Impediment to Development. Westview Press, Boulder, Colorado.

Joarder, Golam Kauser, M. Sayedul Huq, Maksuda Khatun and A.K. Fazlul Huq, 1980, "Studies on the Incidence of Fungal Flora and Deterioration of Aflatoxin in Rice in Bangladesh." Unpublished paper presented at the Post-Production Workshop on Food Grains. Food Science and Technology Division, Bangladesh Council of Scientific and Industrial Research, Dhaka.

Kabir, Khushi, Ayesha Abed and Marty Chen, 1977, "Rural Women in Bangladesh: Exploding Some Myths." Role of Women in Socio-Economic Development in Bangladesh. Proceedings of a Seminar held in Dhaka. Bangladesh Economic Association, Dhaka. pp.72-79.

Karim, A.W. Anwarul and Md.Harunar Rashid, 1979. Extent of Loss of Boro Paddy during Postharvest Operation: A Study Conducted in Bahadurpur Village of Mymensingh District. Agri-Varsity Extension Project, Bangladesh Agricultural University, Mymensingh.

Khan, Azizur Rahman, 1977, "Poverty and Inequality in Rural Bangladesh." Poverty and Landlessness in Rural Asia. ILO, Geneva, pp.137-160.

Khatun, Saleha and Gita Rani, 1977, Bari-Based Postharvest Operations and Livestock Care: Some Observations and Case Studies. Ford Foundation, Dhaka

Kissinger, H., 1975, "Speech to the 7th Special Session of the UN General Assembly." September 1st cited in Bourne, Malcolm C. op. cit.

Krishna, Raj, 1975, "Measurement of the Direct and Indirect Employment Effects of Agricultural Growth with Technical Change." Lloyd G. Reynolds ed., Agriculture in Development Theory, Yale University Press, London, pp.297-326.

Krishnamachari, K.A.V.R., Ramesh V. Bhat, V. Nagarajan and T.B.G. Tilak, 1975, "Hepatitus Due to Aflatoxicosis: An Outbreak in Western India." The LANCET, May 10.

Lal, Deepak, 1972, Wells and Welfare: An Exploratory Cost-Benefit Study of the Economics of Small-scale Irrigation in Maharashtra. Development Centre of the Organisation for Economic Co-operation and Development, Paris.

Lipton, Michael, 1968, "The Theory of the Optimising Peasant." Journal of Development Studies, vol.4, no.3, pp.327-351.

__________, 1977, Why Poor People Stay Poor. Temple Smith Ltd, London.

__________, 1980, "Family, Fungibility and Formality: Rural Advantages of Informal Non-farm Enterprises versus the Urban-formal State." Paper presented at a Conference of the International Economics Association, Mexico

__________, 1983a, "Poverty, Undernutrition, and Hunger." World Bank Staff Working Paper, no.597, Washington D.C.

__________, 1983b, "Labour and Poverty." World Bank Staff Working Paper, no.616, Washington, D.C.

Little, I.M.D. and J.A. Mirrlees, 1974, Project appraisal and planning for developing countries. Heinemann Educational Books,

Lockwood, Merrick, L., 1975a, "Solar Cabinet Dryers for Drying Grain and other Agricultural Produce in Bangladesh: Preliminary Report." International

Voluntary Services and Christian Organisation for Relief and Rehabilitation. Grain Storage Project, Dhaka.

__________, 1975b, "Small Scale Storage and Drying of Paddy in Bangladesh: The Scope for Reducing Losses." Appropriate Agricultural Technology Cell, Bangladesh Agricultural Research Council, Dhaka.

__________, 1977, "Choice of Technology for Rice Processing in Bangladesh." Appropriate Agricultural Technology Cell, Bangladesh Agricultural Research Council, Dhaka.

__________, 1978, "Research and Development Needs for Strengthening the Rice Processing Sector in Bangladesh." Paper presented to the seminar on Food Preservation and Product Development. Bangladesh Council of Scientific and Industrial Research, Dhaka.

__________, 1979a, "Report on a Visit to Some Rice Processing Research and Development Centres of the Region." Appropriate Agricultural Technology Cell, Bangladesh Agricultural Research Council, Dhaka.

__________, 1979b, "The Rubber Roll Rice Huller in Bangladesh." Appropriate Agricultural Technology Cell, Bangladesh Agricultural Research Council, Dhaka.

__________, 1979c, "Current Postharvest Technology Projects in Bangladesh." Appropriate Agricultural Technology Cell, Bangladesh Agricultural Research Council, Dhaka, AATC Information Bulletin no.8.

__________, 1981, "Development and Testing of a Portable Rice Huller for Bangladesh." The Asia Foundation, Dhaka.

Longhurst, Richard, 1980, "Work, Nutrition and Child Malnutrition in a Northern Nigerian Village." Unpublished D.Phil. thesis, University of Sussex

Mahalanobis, P.C., 1961, <u>Experiments in Statistical Sampling in the Indian Statistical Institute</u>. Asia Publishing House, Bombay.

Mahmood, Abu, 1980, "Second Five Year Plan: Approach, Strategy and Techniques; back to Ayub Model." In Qazi Khaliquzzaman Ahmed, ed., <u>Development Planning in Bangladesh - A Review of the Draft Second Five Year Plan</u>, Bangladesh Unnayan Parishad, Special Publication - I, pp.1-25.

Manalo, Eugenio, B., n.d., <u>Agro-Climatic Survey of Bangladesh</u>. Bangladesh Rice Research Institute and International Rice Research Institute.

Mannan, Mahboobul and Burhan Mahmood, 1978, "Optimum Heat Requirement for the Parboiling of Paddy". Unpublished

B.Sc. Engineering thesis, Bangladesh University of Engineering and Technology, Dhaka.

Marx, Karl, 1970, Capital. Vol.1, Lawrence and Wishart, London.

Maxwell, Simon, 1982, "Harvest and Postharvest in Farming Systems Research." IDS Bulletin, vol.13, no.3, pp.21-31.

McCarthy, Florence E., "The Status and Condition of Rural Women in Bangladesh." Mimeo, Dhaka.

Miracle, M.P., 1966, Maize in Tropical Africa. University of Wisconsin Press, Madison.

Miranda, Armindo, 1980, "Nuptuality in Bangladesh." DERAP Publication no.95 Bergen.

Mohamed, N.H., 1981, "Rice Grain Losses from Kada Area (Malaysia): An Overview vis-a-vis Traditional Storage System." Proceedings of the 4th Annual Workshop on Grains Postharvest Technology, SEARCA, Laguna, Philippines, pp.115-131.

Molla, M.R.I., 1972, "Rice Drying Problem during Rainy Season in Bangladesh." Indian Journal of Agricultural Economics, vol.XXVII, no.2, pp.69-71.

National Academy of Sciences, 1978, Postharvest Food Losses in Developing Countries. National Academy of Sciences, Washington, D.C.

Nicholas, M.S.O., 1981, "Postharvest Food Loss Prevention - An International Pespective." Industry and Environment, vol.4, no.1, pp.2-3.

Nurunnabi, B.I., 1975, "Effect of Parboiling and Storage on the Phytic Acid, Inorganic and Total Phosphorus of Some Varieties of Bangladesh Rice." Bangladesh Journal of Scientific and Industrial Research, vol.X, no.1 & 2, pp.38-43.

Nurunnabi, B.I., Dilruba Yasmeen and M. Mominul Huq, 1975, "Effects of Various Treatments of Paddy on the Thiamine, Riboflavin and Niacin Contents of Husked and Milled Rice." Bangladesh Journal of Scientific and Industrial Research,vol.X, no.3 & 4, pp.210-220.

Nurunnabi, B.I. and Abu Zafar Amanatullah, 1979, "Deteriorative Changes in Fat in Rice Stored as Paddy in Different Indigenous Storage Containers: Rancidity and Formation of Free Fatty Acids as Indices of Qualitative Deterioration." Food Science and Technology Division, Bangladesh Council of Scientific and Industrial Research, Dhaka. Unpublished paper.

__________, 1980a, "Deteriorative Changes in Fat in Different Varieties of Bangladeshi Rice During Farm-Level Storage." Paper presented at the

Post-Production Workshop on Food Grain. Food Science and Technology Division, Bangladesh Council of Scientific and Industrial Research, Dhaka.

__________, 1980b, "The Relationship Between Price and Deteriorative Changes in Fat in Rice Procured from the Free Markets in Bangladesh." Paper presented at the Post-Production Workshop on Food Grains. Food Science & Technology Division, Bangladesh Council of Scientific and Industrial Research, Dhaka.

__________, 1980c, "Rancidity in Rice During Village-level Storage." Paper presented at the Post-Production Workshop on Food Grains. Food Science & Technology Division, Bangladesh Council of Scientific and Industrial Research, Dhaka.

Osmani, S.R., 1980, "Poverty, Inequality and the Problem of Nutrition in Bangladesh." Paper presented at a Special Seminar on Food Policy and Development Strategy in Bangladesh. Bangladesh Economic Association, Dhaka.

de Padua, 1976, "Rice Post-Production Handling and Processing: Its Significance to Agricultural Development." Paper presented at the International Workshop on Accelerating Agricultural Development, SEARCA, Laguna, Philippines, April.

Parpia, H.A.B., 1977, "More Than Food Would Be Saved." CERES vol.10, no.6, pp.19-24.

Peers, F.G., C.A. Gilman and C.A. Linsell, 1976, "Dietary Aflatoxins and Human Liver Cancer: A Study in Swaziland." International Journal of Cancer, vol.17, pp.167-176.

Proctor, D.L., and J.Q. Rowley, 1983, "The Thousand Grain Mass (TGM) Method: A Basis for Better Assessment of Weight Losses on Stored Grain" Tropical Stored Products Information, no.45, pp.19-23.

Quasem, Md. Abdul, (1) 1979, "Marketing of Aman paddy with Special Reference to Government Procurement Programmes in Two Selected Areas of Bangladesh." Bangladesh Journal of Agricultural Economics, vol.2 no.I, pp.51-74.

Quasem, Md. Abdul (2) 1980, "Natural Draught Brochure." Rice Post Harvest Technology Project, Bangladesh Rice Research Institute. Unpublished Paper.

Rahman, A., 1970, "Maximal Growth with Wage-Dependent Production Functions." Robinson, E.A.G. and M. Kidron, eds., Economic Development in South Asia. Macmillan, London, pp.379-385.

Raman, C.P., K.S. Narasimhan, B.R. Birewar and G.K. Girish, 1976, "Improvements of Existing Rural Grain Storage

Structures with Special Reference to Andhra Pradesh." Paper presented to the annual Conference of the Indian Society of Agricultural Engineering, Hyderabad.

Ray, Anandarup, 1984, Cost-Benefit Analysis: Issues and Methodologies. A World Bank Publication, The Johns Hopkins University Press, London

Razzaque, M.A., 1980, "Farm Level Storage of Wheat Seeds." Paper presented at the Post-Production Workshop on Food Grains. Food Science & Technology Division, Bangladesh Council of Scientific and Industrial Research, Dhaka.

Russell, Donald G, 1980, "Socio-Economic Evaluation of Grains Post-Production Loss-Reducing Systems in Southeast Asia." Unpublished paper presented at the E.C. Stakman Commemorative Symposium Assessment of Losses which Constrain Production and Crop Improvement in Agriculture and Forestry, University of Minnesota, Minneapolis.

Sahlins, Marshall, 1972, Stone Age Economics, Aldine-Atherton, Chicago.

Salahuddin, Khaleda, 1980, "Women in Bangladesh and the Draft Second Five Year Plan (1980-85)." Qazi Khaliquzzaman Ahmed ed., Development Planning in Bangladesh - A Review of the Draft Second Five Year Plan, Bangladesh Unnayan Parishad, Special Publication - I., pp.106-125.

Samajapati, J.N., M.S. Rahman and T.C. Chowdhury, 1978, Comparative Study of Low Cost Grain Storage Structures for Domestic Use in Bangladesh. Bangladesh Agricultural University, Mymensingh.

Samson, B.T. and B.P.H Duff., 1973, "The Pattern and Magnitude of Field Grain Losses." IRRI Saturday Seminar, July 7th.

Sarker, Md. Rafiqul Islam, 1976, "An Assessment of Postharvest Losses of Rice Grain and Development of a Suitable Method of Drying Rice for Farmers of Bangladesh." A research proposal submitted to the Bangladesh Agricultural Research Council from the Dept. of Farm Power and Machinery, Bangladesh Agricultural University, Mymensingh.

__________, 1980, "Drying of Wet Season Farm Crops in Bangladesh: Some Ideas on the Appropriate Small-Scale Drier." Paper presented at the Post-Production Workshop on Food Grains. Food Science and Technology Division. Bangladesh Council of Scientific and Industrial Research, Dhaka.

Satake, Robert S, 1978, "Status of the Rice Milling Sector." Agricultural Mechanization in Asia, vol.IX, no.2, pp.40-44.

Schultz, T., 1964, Transforming Traditional Agriculture. Yale University Press, New Haven.

__________, 1967, "Significance of India's 1918-19 Losses of Agricultural Labour: A Reply." Economic Journal, vol.77.

Scitovsky, Tibor, 1976, The Joyless Economy: An Enquiry into Human Satisfaction and Consumer Dissastisfaction. Oxford University Press, London.

Scott, Gloria, L., and Marilyn Carr, 1985, "The Impact of Technology on Rural Women in Bangladesh." World Bank Staff Working Paper, no.731, Washington, D.C.

Scott, M.F.G., J.D. MacArthur and D.M.G. Newbery, 1976, Project Appraisal in Practice. Heinemann Educational Books, London.

SEARCA, 1983, Postharvest Quarterly - vol.VIII, No.1, published by the Southeast Asia Cooperative Postharvest Research and Development Programme, Southeast Asian Regional Centre for Graduate Study and Research in Agriculture, Laguna, Philippines.

Sen, A.K., 1972, Choice of Techniques: An Aspect of the Theory of Planned Economic Development. Basil Blackwell: Oxford.

__________, 1975, Employment, Technology and Development. Clarendon Press, Oxford.

__________, 1984, Resources, Values and Development Basil Blackwell, Oxford.

Shahjahan, M., 1975, "Some Aspects of the Ecology and Control of Sitotroga Cerealella Oliv in Bangladesh." Journal of Stored Products Research, vol.11 no.3/4, pp.

Shahnoor, Malik Md., 1980, "Socio-Economics of Rice Milling." Unpublished paper presented at the Post Production Workshop on Food Grains. Food Science and Technology Division, Bangladesh Council of Scientific and Industrial Research, Dhaka.

Shahriar, Latif Md., 1980, "Food Losses During Rice Husking." ADAB News vol.16, November, pp.16-17

Snedecor, George W., and William G. Cochran, 1980, Statistical Methods. The Iowa State University Press, Ames, Iowa.

Spencer, Dunstan S.C., Ibi I. May-Parker and Frank S. Rose, 1976, Employment, Efficiency and Income in the Rice Processing Industry of Sierra Leone. Michigan State University, East Lansing.

Spensley, Philip, 1977, "An Eye on the Granary." CERES, no.60, November-December, pp.15-18.

Spurgeon, David, 1976, Hidden Harvest: A Systems Approach to Postharvest Technology. IDRC, Ottawa, Canada.

Squire, Lyn and Herman G. van der Tak, 1975, Economic Analysis of Projects. The John Hopkins University Press, Baltimore and London.

Srinivasan, S., 1972, "Studies on Degree of Infestation by Rice Weevil Sitophilus Oryzae L. in Different Varieties of Paddy and Rice." Unpublished M.Sc. thesis, Agricultural College, Bapatla.

Sriramulu, M., 1973, "Studies on the Varietal Resistance of Paddy to Lesser Grain Borer, Rhizopertha Dominica (Fab)." Unpublished M.Sc. thesis, Agricultural College, Bapatla.

Stewart, Frances, 1978, Technology and Underdevelopment. The Macmillan Press Ltd, London.

Sukhatme, P.V., 1972, Presidential Address - "Protein Strategy and Agricultural Development." Indian Journal of Agricultural Economics, vol.XXVII, no.1, pp.1-24.

Tickner, Vincent, 1974, "The Rice Processing and Drying Sector and Governments Procurements and Distribution Policies in Bangladesh." Unpublished paper, Dhaka.

Timmer, Peter C., 1974, "Choice of Technique in Rice Milling on Java." Agricultural Development Council, Research and Training Network reprint from Bulletin of Indonesian Economic Studies, vol.2, July, 1973.

Tyler, P.S., 1981, "Reducing post-harvest grain losses by improving the traditional technology." Industry and Environment, vol.4, no.1, pp.14-16.

Tyler, Peter S. and Robin A. Boxall, 1984, "Postharvest Loss Reduction Programmes: A Decade of Activities - What Consequences." Tropical Stored Products Information, No.50, pp.4-13.

UNEP, 1983, Guidelines for Postharvest Food Loss Reduction Activities. United Nations Environment Programme, Industry and Environment Office.

United Nations Industrial Development Organization, 1972, Guidelines for Project Evaluation. United Nations, New York.

United Nations Industrial Development Organization, 1978, Guide to Practical Project Appraisal: Social Benefit-Cost Analysis in Developing Countries. United Nations, New York.

Upton, M., 1962, "Costs of Maize Storage 1959-60 and Cost of Guinea Corn Storage in Silos." Annual Report of West

African Stored Products Research Unit, Nigerian Federal Ministry of Commerce and Industry, Lagos, p.83-85 and 89-90.

USDA, 1980, Rice Postharvest Lossses in Developing Countries United States Department of Agriculture, Science and Education Administration, Agricultural Reviews and Manuals, ARM-W-12.

US Government, 1976, "Commodity Storage Conditions in Bangladesh." A Staff Report to the subcommittee on Foreign Assistance of the Committee on Foreign Relations, United States Senate.

von Harder, Gurdun M., 1975, "Women's Role in Rice Processing." Women for Women. University Press Ltd, Dhaka, pp.66-80.

von Oppen, M., 1976, "Consumer Preferences for Evident Quality Characters of Sorghum." Paper presented at the Symposium on production, processing and Utilization of Maize, Sorghum and Millets. Central Food Technological Research Institute, Mysore.

Ward, Sadie, 1982, Seasons of Change: Rural Life in Victorian and Edwardian England. George Allen & Unwin, London.

Wellisz, S. 1977, "Savings in Isolation and under a Collective Decision Rule" Quarterly Journal of Economics, vol.91, pp.663-666

Westergaard, Kirsten, 1983, Pauperization and Rural Women in Bangladesh: A Case Study. Bangladesh Academy for Rural Development, Comilla

Yates, F., 1946, "A Review of Recent Statistical Developments in Sampling and Sampling Surveys." The Journal of the Royal Statistical Society, A.I09, pp.12-43.

Index